A NEW COSMOLOGY FOR OUR TIME; AN OUTLINE OF THE PRINCIPLES OF TACHYONIC PHYSICS

faster than light; ideas rejected in the twentieth century but not in the twenty first; moving towards the stars and beyond twentieth century dogmas

Malcolm H Sutcliffe PhD

Lulu Publishing Services rev. date: 05/28/2019

TABLE OF CONTENTS

CHAPTER 1

SOME OBSERVATIONS CONCERNING VELOCITIES GREATER THAN THE VELOCITY OF LIGHT AND THE DERIVATION OF THE TACKYONIC LORENTZ TRANSFORMATION EQUATIONS AND THEIR CONVERSE

CHAPTER 2 *(Reproduction of 1994 paper)*

FURTHER OBSERVATIONS CONCERNING VELOCITIES GREATER THAN THE VELOCITY OF LIGHT

CHAPTER 3 *(Reproduction of 1996 PAPER)*

THE PHYSICAL SIGNIFICANCE OF THE IMAGINIFICATION TRANSFOR-MATIONS, THE PRESENTATION OF THE MATRIX ASSOCIATED WITH THE TACKYONIC TRANSFORMATION AND OTHER CONSIDERATIONS

CHAPTER 4

THE "INTERNAL PERIODICAL PHENOMENON" OF LOUIS DE BROGLI,
IS IT THE TRANSFORMATION OF MATTER FROM MINKOWSKI 4 SPACE
TO IDH SPACE AN UNPUBLISHED PAPER FROM MAY 1991

CHAPTER 5

THE ANALOGY OF THE LORENTZ TRANSFORMATION IN IDH SPACE,
IS IT THE UNITARY TRANSFORMATION?

CHAPTER 6

CHAPTER 7

A NEW COSMOLOGY FOR OUR TIME

CHAPTER 8

DISCUSSION OF BELL'S THEOREMS

CHAPTER 9

DISCUSSION OF ASPECT'S EXPERIMENTS AND PROPOSALS FOR FURTHER RELATED EXPERIMENTS

CHAPTER 10

DISCUSSION ON TACHYONIC MECHANICS TWO ALTERNATIVE FORMULATIONS ONE OBTAINED BY THE AUTHOR INDEPENDENTLY OF THE EARLIER VERSION

CHAPTER 11

A SPECULATION AS TO THE NATURE OF DARK ENERGY AND DARK MATTER, TACHYONIC GRAVITY.

CHAPTER 12

A NOTE ON NOMENCLATURE

The author uses both terms for faster than light objects ie. both "tachyonic" his preferred term and "tackyonic". Because CHAPTER THREE is a reproduction of a previously published paper the term tackor is used for a faster than light vector, this could equally be termed a tachor. In a second expanded edition of this book, this term will be used, but a reader should not be confused by these minor spelling differences, which occur in the text at various points.

Chapter 1

SOME OBSERVATIONS CONCERNING VELOCITIES GREATER THAN THE VELOCITY OF LIGHT AND THE DERIVATION OF THE TACKYONIC LORENTZ TRANSFORMATION EQUATIONS AND THEIR CONVERSE

SECTION I INTRODUCTION. THE REFUTATION OF THE 1907 EINSTEIN PROOF

Although is has become customary to believe that it is not possible for particles to travel faster than c, the velocity of light, the proof of this has never been effectively provided. Such particles which travel faster than the velocity of light have come to be known as tachyons. The origin being the Greek word 'tachys' which means fast as in fast velocity transcribed from the Greek.

Although no effective proof has been provided it is instructive to look at the original Einstein proof in order to learn about the basis of tackyonic physics. The proof was provided in 1907 A. Einstein[1] references as at the end of the chapter. The basis of the proof we will reproduce here for our education in order to see the fundamental flaw in this out dated mode of thinking.

What Einstein said was suppose that an effect in a system K (inertial system) is propagated with velocity $U > C$ but suppose that another system K^1 is moving with velocity $V < C$ relative to K. Then *if we choose*

$$\frac{C}{\boldsymbol{u}} < \frac{V}{C} < 1 \dots (A) \qquad U^1 = \frac{U - V}{1 - \frac{UV}{C^2}} < 0$$

from this we arrive at the following contradiction. That if in K an event A (with velocity U) causes an event B (with velocity U) then in K^1 the event B causes A (an account of the negative nature of U^1). From this argument, it is then *assumed* that no *information* can be transmitted beyond the velocity of light. The *assumption* being that both energy and information both have to be transmitted at velocities less than or equal to the velocity of light. The mistake in my humble opinion that authors make is in not distinguishing between information which I will later emphasis is transmitted when ever a probability distribution is transmitted and mathematically real energy. It is the view of the author that real energy can be transformed into mathematically imaginary energy by a process known as imaginification and that this imaginified energy can carry information faster than the velocity of light. It is also the view of the author that matter having gone through the process of imaginification can also go through the reverse process of changing from mathematically imaginary matter into ordinary tardonic matter. We use the term tardonic from the root tard which means slow, a tardon being a particle that travels at velocities less than C.

Before returning to the above remarks I'd like to point to in an unproven assumption in the above Einstein 'proof'. Namely that having chosen V, say we have the freedom to choose U so as to fullfill condition (A)

ie. $\frac{C}{U} < \overset{\text{v}}{\frac{}{C}} < 1.$ condition (A)

Let us proceed instead by assuming that $U' \geq 0$ and see where it leads us.

$$U' \geq 0$$

$$\Rightarrow \frac{U - V}{1 - \frac{UV}{C^2}} \geq 0.$$

However since according to the assumptions in Einstein $U > C$ and $V < C$

$$U - V > 0.$$

Thus

$$\frac{1}{1 - \frac{UV}{C^2}} \geq 0$$

$$\Rightarrow 1 - \frac{UV}{C^2} \geq 0 \Rightarrow \frac{UV}{C^2} \leq 1$$

or

$$U \leq \frac{C^2}{V}.$$

However, in the limit $V \to C$ if $U < \frac{C^2}{V}$ then U would become less than C, thereby creating a contradiction in the limit.

In other words the unproven assumption is that there is no relationship between U and V, or the relationship that does exist always satifies (A).

Thus the relationship $U = \frac{C^2}{V}$ is apparently of importance. And clearly if $U = \frac{C^2}{V}$ then Einsteins objection that $U' < 0$ and his concern about an alleged contradiction is no longer valid. However there is another fundamental *mathematical* objection to the Einstein proof which I am surprised other authors have not pointed to with greater regularity. The mathematical objection is as follows. As we have seen the Einstein proof employs the special relativistic law for the sum and difference of velocities ie. $U' = \frac{U - V}{1 - \frac{UV}{C^2}}$. This law is derived using Lorentz transformations. Thus the proof is dependent on Lorentz transformations.

2

In Section II I will prove rigourously that the Lorentz transformations imply that matter traveling tackyonically will inevitably have imaginary space-time coordinates. Einstein talks about an event A being before an event B in time. However mathematically imaginary numbers or complex numbers can not be ordered unambiguously in the manner Einstein suggests. This is a well known result from complex analysis. No complex or imaginary number can be said unambiguously to be greater or less than another which is precisely the mistake contained in the flawed proof which the author[2] first contradicted personally in public on Sept 18th 1991 in a plenary session of the Russian Academy of Sciences although he had written an unpublished paper on this subject as early as 1987. Many of the ideas of this book were known to the author in the late 1980s and the book is based on the unpublished papers from that time, advances made for the preparation of the 1991 paper and further advances made since 1996.

From a historical viewpoint it is interesting to recall that the velocity $\frac{C^2}{V}$ which the author chooses to describe as the universal corresponding velocity V_c was first given by M.L. de Broglie[3] and he assumed it to be the phase velocity of a wave which "couldn't carry energy according to Einstein's ideas". This is a comment that is correct but M.L. de Broglie did not appreciate the full significance of the velocity $\frac{C^2}{V}$ merely using it as a device to obtain his now famous expressions for the equations of wavelength and frequency of matter particles. However M.L. de Broglie considered the velocity V_c not as a fundamental property of space-time but as a tool relevant only to the quantum level. The author by contrast derives $V_c = \frac{C^2}{V}$ in Section III without using quantum assumptions. In chapter three the author will show how by using the ideas of chapter one it is possible to reinterpret much of quantum mechanics in a new light using tackyonic concepts. This new light will enable us to more properly understand the nature of physical reality much of which is hidden from us because of its tackyonic nature.

P.A.M. Dirac[4] also derived the velocity $\frac{C^2}{V}$ in a quantum mechanical manner using the approach to quantum mechanics he developed (that of the ket vector, bra vector, state vector etc) which the author finds most illuminating. He considered it to be the phase velocity of a wave associated with a free particle which had group velocity V. This wave associated with the free particle was the state vector of the particle. Thus P.A.M. Dirac's wave had taken on extra meaning from the original L. de Broglie idea but both were seen as purely quantum mechanical phenomena.

The author finds it interesting to note that there is nothing in the definition of information to restrict the transmission of information to velocities less than C.

The mathematical definition of information according to A.I. Khinchin[5] $= \sum_{K=1}^{n} PK \log PK$ where PK is a probability. This makes plain that as long as (effectively) probabilities are transmitted down 'channels' then information is transmitted. Although it has been customary to assume such channels are

3

composed of real energy at speeds less than C or at C <u>nothing</u> in the <u>definition</u> of information makes this assumption an a priori necessity.

SECTION II. THE PROOF THAT THE LORENTZ TRANSFORMATIONS RESULT IN IMAGINARY SPACE-TIME CO-ORDINATES FOR TACKYONIC VELOCITIES

We have the Lorentz transformations that are well known to mathematical physicists all over the world. The transformation being employed here being the restricted transformation applying to the special case of two inertial frames K and K^1 which move along the common x-axis with velocity V in the usual fashion

supppose $V = (1 + \theta)C$ $\qquad \theta > 0$ and wholly real.

then if we substitute into

$$X^1 = \frac{x - vt}{V1 - \frac{V^2}{C^2}}$$ where we assume x, t to be tardonic and real

let $\beta = \dfrac{1}{\sqrt{1 - \frac{V^2}{C^2}}} =$

Suppose $V > C$ i.e. $V = (1 + \theta)C$ $\qquad \theta > 0$ $\qquad \theta$ wholly real

let

$$\beta = \frac{1}{\sqrt{1 - \frac{v^2}{c^2}}} \qquad = \frac{1}{\sqrt{1 - \frac{(1+\theta)^2 c^2}{c^2}}} \qquad = \frac{1}{\sqrt{1 - (1+\theta)^2}}$$

$$= \frac{1}{\sqrt{1 - 1 - 2\theta - \theta^2}} \qquad = \frac{1}{\sqrt{-2\theta - \theta^2}} \qquad = \frac{1}{\sqrt{-1}\sqrt{2\theta + \theta^2}}$$

$$= \frac{1}{i\sqrt{2\theta + \theta^2}} \qquad = \frac{1}{iK} \qquad\qquad K = \sqrt{2\theta + \theta^2}$$

$$= \frac{-i}{K}$$

to prove

$$x - Vt < 0 \qquad V > 0$$
$$x - Vt < 0 \text{ iff } \tfrac{x}{t} < V \quad \text{by assumption}$$
$$\text{true} \quad v > C$$

thus

$$x - Vt < 0$$
$$\text{suppose} \quad x - Vt = -Q \quad Q > 0$$

thus

$$x^1 = \frac{iQ}{K}$$

conversely if

$$x^1 = \frac{iQ}{K} \quad \begin{array}{l} Q > 0 \\ K > 0 \end{array} \qquad \text{then V>0}$$

4

i.e. x^1 is wholly imaginary with a positive coefficient. In the above **proof that** tackyonic nature is mathematically imaginary **we make an assumption.** Namely **that** Lorentz transformations are valid for the description of tackyonic matter which is an assumption that the author makes as a working hypothesis to see where it may lead us, then co-ordinates which are mathematically imaginary are travelling at tackyonic velocities. We know from above that subject to the stated assumption that tackyonic velocities lead to imaginary co-ordinates but do imaginary co-ordinates imply tackyonic velocities?

Let x^1 be a positive space imaginary co-ordinate ie. $x^1 = \frac{ai}{b}$ equation we assume that the unprimed co-ordinates are travelling in the existing positivity of real energy which we all know through direct experience, ie. those co-ordinates that form what is known as an inertial frame of reference in special relativity.

Next we prove the corresponding result for time co-ordinates. Suppose V> C ie. $V = (1+\theta)C \ \theta > 0 \ \theta$ wholly real

$$\beta = \frac{1}{\sqrt{1 - \frac{v^2}{c^2}}} = \frac{1}{\sqrt{1 - \frac{(1+\theta)^2}{c^2}}} = \frac{1}{\sqrt{1 - (1+\theta)^2}}$$

$$= \frac{1}{Ki} \qquad K = \pm\sqrt{2\theta + \theta^2}$$

in fact it is not necessary to choose the $-ve$ value for K as a simple proof shows (on the strict assumption $V > C$) that

$$t - \frac{vx}{c^2} \text{<0} \quad t^1 = (t - \frac{Vx}{c^2}) \ \frac{1}{\beta}). \quad \textsf{(V=v)} \text{ in this proof}$$

Proof

$$t - \frac{vx}{c^2} < 0 \ (A) \ -: \text{ suppose } t - \frac{vx}{c^2} = -p \qquad p > 0$$

$$\text{iff } \frac{c^2}{v} > \frac{x}{t}$$

clearly $\frac{x}{t} < V$ [by sandwich principle of inequalities
$$A < B < C \text{ iff } A < C]$$

iff $\frac{c^2}{v} < V$ since this is true by assumption $V > C$

the original assumption (A) is true.

Thus we have that $t^1 = \frac{iP}{K}$ where $P > 0 \ K > 0$.
This we have theorem 1 section 2 which can be stated as: - travel beyond the velocity of light is of an imaginary nature in time and space.

Next we prove the counter-result that imaginary space-time is beyond C

i.e. $x^1 = \dfrac{ai}{b}$ where $\begin{array}{l} a \geq 0 \quad \text{wholly real} \\ b \geq 0 \quad \text{wholly real} \end{array}$

let $a = Vt - x$

thus $Vt - x > 0$ for all x, V, t.

if $Vt > x$

if $\dfrac{x}{t} < V$. Suppose $\dfrac{x}{t} \to C$ then if $\dfrac{x}{t} < V$ then $V > C$ (ignoring the case $\dfrac{x}{t} = C$ which leads to the statement $C = C$). This implies that $a = Vt - x > 0$ means $V > C$ thus $-a = x - Vt (i = \sqrt{-1})$ as always.

Consider $\dfrac{i}{-b} = \dfrac{-i}{b}$ (in order $\dfrac{-i}{b} - a = \dfrac{ia}{b}$ as above).

$$\dfrac{-i}{b} = \dfrac{1}{bi}.$$

Let $b = \sqrt{2\theta + \theta^2} : \theta > 0$ otherwise b not necessarily $+Ve$ real. (A)

$$\dfrac{1}{bi} = \dfrac{1}{(\sqrt{2\theta + \theta^2})i}$$

$$= \dfrac{1}{\sqrt{-1}\sqrt{2\theta + \theta^2}} = \dfrac{1}{\sqrt{-2\theta - \theta^2}} = \dfrac{1}{\sqrt{1 - 1 - 2\theta - \theta^2}}$$

$$= \dfrac{1}{\sqrt{1 - (1 + \theta)^2}} = \dfrac{1}{\sqrt{1 - (1 + \theta)^2 \frac{C^2}{C^2}}} \quad \text{let} \quad V = (1 + \theta)C \quad \theta > 0 \quad \text{by} \quad (A)$$

$$\dfrac{1}{bi} = \dfrac{1}{\sqrt{1 - \frac{V^2}{C^2}}} \quad V > C \quad \text{Q.E.D.}$$

A similar proof demonstrates that imaginary time has velocity greater than C.

SECTION III. THE DERIVATION OF THE UNIVERSAL CORRESPONDING VELOCITY V_c

In the prevous section it is shown that matter moving with velocity greater than C, henceforth known as tackyonic matter must possess imaginary co-ordinates in space and time (No comment is made so far on the nature of the tackyonic matter and this section is self contained as far as the nature of tackyonic matter is concerned, i.e. it is not necessary to make any comment on the nature of the tackyonic matter for the purpose of this section.)

If the 'world' of tackyonic matter is to have any physical importance to us there must be some exchange of matter or information about the tackyonic matter available to our manifest (originated from Greek work meaning "at hand") 4-space world.

If the tackyonic matter never interacts in any way with our 4-space it can neither be proved or disproved that it exists; its properties might be interesting to mathematicians but not to engineers or physicists. If an exchange occurs then it could be just one way, i.e. from tardonic matter (the matter in 4-space) to tackyonic or from tackyonic to tardonic matter only, or both processes could occur simultaneously and as converse elements in an ongoing natural phenomenon.

Let us however just consider the transformation of the space time co-ordinates from 4-space into the tackyonic state.

Let us consider an event with respect to a frame of reference K with co-ordinates (x, y, z, t) and similarly with respect to another frame of reference $K^1(x^1, y^1, z^1, t^1)$ moving with velocity V relative to K along the positive x-axis.

Let us now propose that the event undergoes the above transformation in the frame of reference K^1. This transformation the author has given the name "imaginification" for reasons that will soon be self evident. The question that arises is how do we describe this transformation mathematically?

A clue is given by the theorems of the previous section. If matter becomes tackyonic its space time description is imaginary. So let us multiply x^1 and t^1 by i where $i^2 = -1$. We thus have two quantities ix^1 and it^1 which we known from the previous section are tackyonic.

The question that arises now is what velocity are these imaginary quantities moving relative to K we call this velocity the universal corresponding velocity. (Also that velocity relative to K^1; the imaginification velocity VI). The term "corresponding" because it is the tackyonic velocity that corresponds to a particular tardonic velocity V, the "universal" because this velocity is a fundamental time space relationship that is arrived at in a very simple but general fashion without any restricting assumptions.

$$x^1 = \frac{x - Vt}{\sqrt{1 - \frac{V^2}{C^2}}} \dots (1) \qquad t^1 = \frac{t - \frac{Vx}{C^2}}{\sqrt{1 - \frac{V^2}{C^2}}} \dots (2)$$

Now we apply the imaginification transformation to the 'event' (x^1, t^1) noting that the velocity V has transformed into V_C the universal corresponding velocity.

$$ix^1 = \frac{x - V_C t}{\sqrt{1 - \frac{V_C^2}{C^2}}} \dots (3) \qquad it^1 = \frac{t - \frac{V_C x}{C^2}}{\sqrt{1 - \frac{V_C^2}{C^2}}} \dots (4)$$

(We ignore the (y, z) co-ordinates although the trivial relation $y^1 = y$, $z^1 = z$ implies $iy^1 = iy$ and $iz^1 = iz$).

7

Also from (1)

$$ix^1 = \frac{i(x - Vt)}{\sqrt{1 - \frac{V^2}{C^2}}} = \frac{Vt - x}{\sqrt{\frac{V^2}{C^2} - 1}} \quad \cdots (5)$$

and from (2)

$$it^1 = \frac{i[t - \frac{Vx}{C^2}]}{\sqrt{1 - \frac{V^2}{C^2}}} = \frac{\frac{Vx}{C^2} - t}{\sqrt{\frac{V^2}{C^2} - 1}} \quad \cdots (6)$$

equate (5) to (3) and (4) to (6)

$$\frac{Vt - x}{\sqrt{\frac{V^2}{C^2} - 1}} = \frac{x - (V_C)t}{[1 - \frac{(V_c)^2}{C^2}]^{1/2}} \quad \cdots (7)$$

$$\frac{\frac{Vx}{C^2} - t}{[\frac{V^2}{C^2} - 1]^{1/2}} = \frac{t - \frac{(V_c)x}{C^2}}{[1 - \frac{(V_c)^2}{C^2}]^{1/2}} \quad \cdots (8)$$

$$(7) \Rightarrow t\left\{ V\left[1 - \frac{(V_C)^2}{C^2}\right]^{1/2} + V_C \left[\frac{V^2}{C^2} - 1\right]^{1/2} \right\} = x\left\{ \left[\frac{V^2}{C^2} - 1\right]^{1/2} + \left[1 - \frac{(V_C)^2}{C^2}\right]^{1/2} \right\}$$

$$(8) \Rightarrow t\left\{ \left[\frac{V^2}{C^2} - 1\right]^{1/2} + \left[1 - \frac{V_C^2}{C^2}\right]^{1/2} \right\} = x\left\{ \frac{V}{C^2}\left[1 - \frac{V_C^2}{C^2}\right]^{1/2} + \frac{V_C}{C^2}\left[\frac{V^2}{C^2} - 1\right]^{1/2} \right\}$$

Dividing we obtain the following equation that has eliminated x^1, t^1, x and t.

$$\left\{ \left[\frac{V^2}{C^2} - 1\right]^{1/2} + \left[1 - \frac{(V_C)^2}{C^2}\right]^{1/2} \right\}^2 = \left\{ V\left[1 - \frac{(V_C)^2}{C^2}\right]^{1/2} + V_C\left[\frac{V^2}{C^2} - 1\right]^{1/2} \right\} \times A$$

$$A = \left\{ \frac{V}{C^2}\left[1 - \frac{(V_C)^2}{C^2}\right]^{1/2} + \frac{V_C}{C^2}\left[\frac{V^2}{C^2} - 1\right]^{1/2} \right\}$$

We equate coefficients of the Square Root Term

$$2 = \frac{VV_C}{C^2} + \frac{VV_C}{C^2} \qquad \text{SRT is} \quad \left[\frac{V^2}{C^2} - 1\right]^{1/2}\left[1 - \frac{(V_C)^2}{C^2}\right]^{1/2}$$

Thus we have that

$$1 = \frac{VV_C}{C^2} \Rightarrow V_C = \frac{C^2}{V}.$$

Thus the universal corresponding velocity is $\frac{C^2}{V}$.

$$V_C = \frac{VI + V}{1 + \frac{VIV}{C^2}} \Leftrightarrow VI = \frac{V_C - V}{1 - \frac{V_C V}{C^2}}$$

$$VI = \frac{\frac{C^2}{V} - V}{1 - \frac{C^2 V}{C^2 V}} = \infty.$$

8

Thus if we substitute into (3) and (4) for V_C we obtain the following equations.

$$ix^1 = \frac{x - \frac{C^2}{V}t}{V1 - \frac{C^2}{V^2}} \qquad (9)$$

$$it^1 = \frac{t - \frac{x}{V}}{\sqrt{1 - \frac{C^2}{V^2}}} \qquad (10)$$

These are the tackyonic Lorentz imaginification transformation equations and have been published in Sutcliffe[2] and Sutcliffe[6]. They should be strictly described as the restricted tackyonic Lorentz imaginification transformation equations because they have been derived from the restricted Lorentz transformation equations.

SECTION IV. THE CONVERSE DERIVATION USING THE IDEA OF POSITIVISATION WHICH IS THE REVERSE PROCESS TO IMAGINIFICATION

Suppose we have a tackyonic particle with tackyonic (and thus from Section II imaginary) co-ordinates say ix' and it'.

In Section III we derived the value for the universal corresponding velocity V_C which is contained in equations (3) and (4). These can be rewritten as the tackyonic Lorentz transformation imaginification equation (9) and (10). This involved deriving the motion in the imaginary space in which it travels. (I will later argue that this space is infinite dimensional and term it infinite-dimensional hyperspace or I.D.H space).

However we propose to reuse equations (3) and (4) so as to obtain V_p the positivisation velocity. The positivisation velocity is the velocity which we would obtain from our tackyonic velocity which we choose to still call V_C. However here because we start in the tackyonic state, it is V_C that is primary, being described by (3) and (4) and our objective is to derive V_p.

The question is what do we do to equation (3) and (4) to get us back to a tardonic (ie slow) state. (tard French for slow)

Clearly if we multiply the L.H.S of (3) and (4) by $\pm i$ we obtain a real quantity. We chose to multiply by $-i_1$ for two reasons. Firstly because it takes us back to our original x^1 and t^1 and secondly because it makes the positivisation process the reverse of the imaginification process in a strictly mathematical manner. In one process we multiply by $+i$ in the reverse process we multiply by its converse $-i$.

As before when we carry out the positivisation process we obtain our new velocity on the other side of the light barrier. (The author suggests this process to occur instantaneously at the sub quantum or quantum level and will discuss the quantum mechanical implications of these ideas later in the book.) This new velocity obtained from our tackyonic V_C we chose to call V_p the positivisation velocity.

Hence we have from (3) and (4) previous section

$$ix' = \frac{x - V_C t}{\sqrt{1 - \frac{V_C^2}{C^2}}} \dots (1) \quad it^1 = \frac{t - \frac{V_C x}{C^2}}{\sqrt{1 - \frac{V_C^2}{C^2}}} \dots (2)$$

Now we apply the positivisation transformation to the 'event' (ix', it') noting that the tackyonic velocity V_C has transformed into V_p the positivisation velocity.

Multiplying the L.H.S by $-i$ and positivising we obtain

$$x^1 = \frac{x - V_p t}{\sqrt{1 - \frac{V_p^2}{C^2}}} \dots (3) \quad t^1 = \frac{t - V_p \frac{x}{C^2}}{\sqrt{1 - \frac{V_p^2}{C^2}}} \dots (4)$$

Thus from equation (1), (2), (3) and (4) we hve (4) equations in C, V_p, x, x^1, t, t^1. We wish to eliminate x, x^1, t, t^1 (which is quite possible with (4) equations) in order to obtain a relationship between V_p, V_C and C (as before in our derivation of the universal corresponding velocity).

Multiply equation (1) and (2) by $-i$ in order to obtain x^1 and t^1

$$(1) \times \text{ by } -i \quad x^1 = \frac{-i(x - V_C t)}{\sqrt{1 - \frac{V_C^2}{C^2}}} = \frac{x - V_C t}{\sqrt{\frac{V_C^2}{C^2} - 1)}} \dots (5)$$

$$(2) \times \text{ by } -i \quad t^1 = \frac{i(t - \frac{V_C x}{C^2})}{\sqrt{1 - \frac{V_C^2}{C^2}}} = \frac{t - V_C \frac{x}{C^2}}{\sqrt{\frac{V_C^2}{C^2} - 1}} \dots (6)$$

equate (5) to (3) and (4) to (6) to eliminate x^1 and t^1

$$\frac{x - V_C t}{\sqrt{\frac{V_C^2}{C^2} - 1}} = \frac{x - V_p t}{\sqrt{1 - \frac{V_p^2}{C^2}}} \dots (7) \quad (5) = (3)$$

$$\frac{t - \frac{V_C x}{C^2}}{\sqrt{\frac{V_C^2}{C^2} - 1}} = \frac{t - V_p \frac{x}{C^2}}{\sqrt{1 - \frac{V_p^2}{C^2}}} \dots (8)$$

$$(7) \Rightarrow t \left\{ V_p \left(\frac{V_C^2}{C^2} - 1 \right)^{\frac{1}{2}} - V_C \left(1 - \frac{V_p^2}{C^2} \right)^{\frac{1}{2}} \right\}^{\frac{1}{2}} = x \left\{ \left(\frac{V_C^2}{C^2} - 1 \right)^{\frac{1}{2}} - \left(1 - \frac{V_p^2}{C^2} \right)^{\frac{1}{2}} \right\}$$

$$(8) \Rightarrow t \left\{ \left(1 - \frac{V_p^2}{C^2} \right)^{\frac{1}{2}} - \left(\frac{V_p^2}{C^2} - 1 \right)^{\frac{1}{2}} \right\} = x \left\{ \frac{V_C}{C^2} \left(1 - \frac{V_p^2}{C^2} \right)^{\frac{1}{2}} - \frac{V_p}{C^2} \left(\frac{V_C^2}{C^2} - 1 \right)^{\frac{1}{2}} \right\}$$

dividing we obtain the following equation (that has eliminated t, x, t^1 and x^1) from (7) and (8)

$$\frac{\left\{ V_p \left(\frac{V_C^2}{C^2} - 1 \right)^{\frac{1}{2}} - V_C \left(1 - \frac{V_p^2}{C^2} \right)^{\frac{1}{2}} \right\}}{\left\{ \left(1 - \frac{V_p^2}{C^2} \right)^{\frac{1}{2}} - \left(\frac{V_C^2}{C^2} - 1 \right)^{\frac{1}{2}} \right\}} = \frac{\left\{ \left(\frac{V_C^2}{C^2} - 1 \right)^{\frac{1}{2}} - \left(1 - \frac{V_p^2}{C^2} \right)^{\frac{1}{2}} \right\}}{\left\{ \frac{V_C}{C^2} \left(1 - \frac{V_p^2}{C^2} \right)^{\frac{1}{2}} - \frac{V_p}{C^2} \left(\frac{V_C^2}{C^2} - 1 \right)^{\frac{1}{2}} \right\}}$$

$$\Rightarrow \left\{ \left(\frac{V_C^2}{C^2} - 1 \right)^{\frac{1}{2}} - \left(1 - \frac{V_p^2}{C^2} \right)^{\frac{1}{2}} \right\}^2 = \left\{ V_p \left(\frac{V_C^2}{C^2} - 1 \right)^{\frac{1}{2}} - V_C \left(1 - \frac{V_p^2}{C^2} \right)^{\frac{1}{2}} \right\} \times A$$

$$A = \left\{ \frac{V_C}{C^2} \left(1 - \frac{V_p^2}{C^2} \right)^{\frac{1}{2}} - \frac{V_p}{C^2} \left(\frac{V_C^2}{C^2} - 1 \right)^{\frac{1}{2}} \right\}$$

We equate the coefficients of the square root term as in our calculation in section III to obtain that

$$-2 = \frac{V_p V_C}{C^2} + \frac{V_p V_C}{C^2}$$

$$-2 = \frac{2 V_p V_C}{C^2} \quad V_p = -\frac{C^2}{V_C}$$

if we substitute

$$V_C = \frac{C^2}{V}$$

we obtain that $V_p = -V$.

Thus by using our chosen definition of positivisation we obtain a negative velocity from the one we started from. What is also clear is that if we had employed negative co-ordinates $(-x, -t)$ on positivisation (using positive values for our imaginary multiplier) we would obtain $V_p = V$.

Thus if we substitute $V_p = -\frac{C^2}{V_C}$ into (3) and (4) we obtain the restricted tackyonic positivisation equations

$$x^1 = \frac{x + \frac{C^2}{V_C} t}{\sqrt{1 - \frac{C^2}{V_C^2}}}$$

$$t^1 = \frac{t + \frac{x}{V_C}}{\sqrt{1 - \frac{C^2}{V_C^2}}}$$

The author explains the negative nature of V_p because the positivisation is being considered from the point of view of the receding tackyonic co-ordinates ie. from the opposite direction. Because it is from the opposite point of view V_p becomes negative. This is shown on FIGURE ONE the next page.

FIGURE ONE SHOWING POSITIVISATION VELOCITY

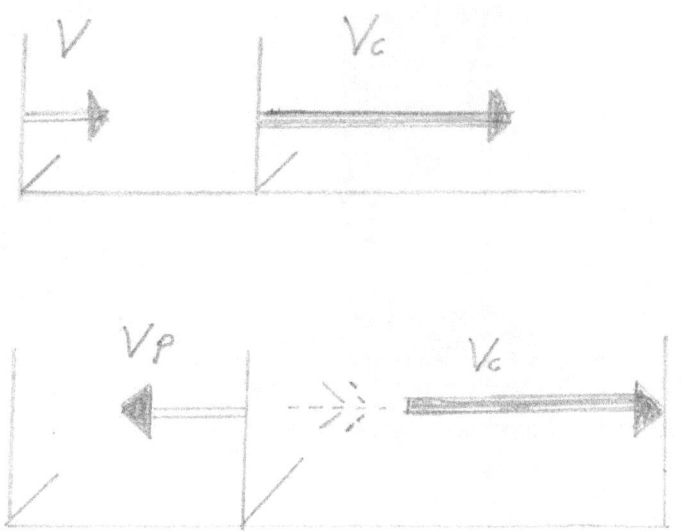

REFERENCES

1. A. Einstein Ann. PHYSICS L17 **23** (1907) p.371.

2. M. Sutcliffe. Proceedings of SEPT 1991, 2nd International Conference on Space Time and Gravitation (Published St. Petersbourg 1994).

3. L de Broglie., Phil. MAG **47** (1924), p.446.

4. PAM DIRAC. The Principles of Quantum Mechanics, 4th Edition, Oxford University Press. Section 32 (1947).

5. A.I. Khinchin. The Entrophy Concept in Probability Theor, MATHEMATICHESKIKH NAUK Vol.VIII, no.3 (1953).

6. M. Sutcliffe. Proceedings 3rd Internal Conference on Space Time and Gravitation held 1994, Published 1995 [reprint of CHAPTER 2].

Chapter 2

FURTHER OBSERVATIONS CONCERNING VELOCITIES GREATER THAN THE VELOCITY OF LIGHT

M.H. Sutcliffe
Physics Dept. UMIST P.O. Box 88 Manchester M60 1QD UK

SECTION 1. THE AUTHOR DESCRIBED THE IMAGINIFICATION TRANSFORMATION AT THE RAS CONFERENCE 1991, ST. PETERSBURG AND THIS WAS PUBLISHED 1994 (PAGES 152-162) BY RAS ST.PETERSBURG

The effect of this transformation is to transform the Lorentz transformations into the Tackyonic Lorentz transformation equations. The velocity V is transformed into the corresponding velocity V_c and the space-time description becomes imaginary. These Tackyonic transformations are (1) and (2) which on substitution for V_c become (3) and (4). In this paper the author will further explain the converse process of positivisation and will make a suggestion that the "internal periodical phenomeon" of Louis de Broglie (1924) is the process of imaginification and the converse process of positivisation acting on a particle at a point. It is suggested that the frequency with which these converse operations act at a point on a mass is the same as L. de Broglie gave for the frequency of the "internal periodical phenomenon". This enables us to make a link between tackyonic space and the space in which the equations of quantum mechanics are framed. The author suggests that tackyonic space which he terms IDH space (infinite dimensional hyper - space) is both holographic and infinite dimensional. He suggests that it is holographic as a consequence of the infinite velocity of communication between "parts" of tackyonic space.

$$ix^1 = \frac{x - V_c t}{\sqrt{1 - V_c^2/c^2}} \tag{1}$$

$$it^1 = \frac{1 - V_c x/c^2}{\sqrt{1 - V_c^2/c^2}}, \quad v_c = \frac{c^2}{v} \tag{2}$$

substituting $V_c = C.C/V$ the universal corresponding velocity derived in Sept. 1991 as above.

$$ix^1 = \frac{x - (c^2/V)t}{\sqrt{1 - c^2/V^2}} \tag{3}$$

$$it^1 = \frac{t - x/V}{\sqrt{1 - c_2/V^2}} \tag{4}$$

Reference L. de Broglie, Phil.Mag.47, (1924), p.446.

SECTION 2. THE DESCRIPTION OF THE POSITIVI-SATION TRANSFORMATION

On imagnification a particle at a point with mass M_0 changes to become imaginary in space and time coordinates and also changes the sign of the mass M_0 to become - M_0 and the sign of its charge changes as explained in section 5 of the 1991 paper. We can define a converse operation of positivisation which transforms our $-M_0$ particle in tackyonic space to become $+M_0$ again in 4-space. The charge will also change sign again on positivisation. We positivise our two imaginary coordinates (ix', it') by multiplying by $-i$ and the two coordinates become (x', t'). The velocity that (x', t') travel relative to (x, t) I call V_p the positivisation velocity. We have four equations (1) and (2) from section 1 and (5) and (6) which we obtain by positivising (1) and (2) as described. If we carry out the algebra in a similar manner to the 1991 derivation of V_p the universal corresponding velocity we obtain by eliminating x, t, x', t' and substituting for V_c that V_p has the value $-V$.

$$ix^1 = \frac{x - V_c^t}{\sqrt{1 - V_c^2/c^2}} \quad (1) \qquad it^1 = \frac{t - V_c x/c^2}{\sqrt{1 - V_c^2/c^2}}, \quad V_p = -c^2/V_c \quad (2)$$

on positivising we obtain

$$x^1 = \frac{x - V_p^t}{\sqrt{1 - V_p^2/c^2}} \tag{5}$$

$$t^1 = \frac{t - V_p x/c^2}{\sqrt{1 - V_p^2/c^2}}, \quad V_c = c^2/V, \quad V_p = -V \tag{6}$$

In addition to the transformation from tackyonic space back to 4-space the positivisation process also changes the rest of mass M_0 back to positive sign and the rest charge density changes sign also the momentum and energy change from imaginary back to real quantities. To summarise the two transformations in a table:

14

IMAGINIFICATION	POSITIVISATION
rest mass: changes to $-M_0$	changes back to $+M_0$
charge density: changes sign	changes sign back
energy, momentum: become iE, iP	change back to E, P
space, time coordinates: have velocity V_c	have velocity V_p described by (5), (6)

The change of sign of the mass from negative to positive justifies the use of the term positivistion. The two transformations are thus converse operations that change from 4-space to tackyonic space and back from tackyonic space to 4-space. A possible reason for the negative sign of V_p is that it is being considered from the point of view of our imaginified frame.

SECTION 3. THE IDEAS OF LOUIS DE BROGLIE

In the paper already quoted L.de Broglie said that the energy of a particle should be described to an "Internal Periodical Phenomenon" or IPP and the frequency of the IPP was to be given by (7) below i.e. the rest mass energy divided by Planks constant. It was an arbitrary assumption to make because at no point in the key paper did he suggest what the IPP was physically.

However if he hadn't made this assumption then he wouldn't have been able to make further progress. He then considered the mass moving with velocity V away from a fixed observer and said that the corresponding frequency to the one in (7) would be given by (8). And he also pointed out that the IPP would have frequency given by (9) if the mass M_0 with IPP given by (7) was considered by a stationary observer. Thus we have three frequencies, the IPP given by (7), the corresponding frequency to the IPP for the fixed observer but moving mass with corresponding energy given by (8a) and frequency (8) and finally the IPP given by a stationary observer in the mass M_0 was considered to move away with velocity V as in (9). He then says that frequencies (8) and (9) stay in phase provided that (8) is the frequency of a wave with velocity V_c where V_c is the same as the one derived by the author *without* making quantum assumptions. This allows the author to make a tentative guess as to what the IPP actually is. I suggest that it is the process of matter particles engaging in imaginification and the converse process. So if there is a particle at (x, t) then the particle will transform into tackyonic space which the author calls IDH space. However simultaneously another particle will replace the now imaginified particle by positivisation, i.e. the M_0 becomes $-M_0$ simultaneously with the positivisation of a $-M_0$ becoming M_0. The frequency of this process is given by (7). The reason why the particle must be replaced by another is because we haven't yet observed the total disappearance of matter. If however it was possible to prevent the replacement then we would have the possibility of sending matter into IDH space and positivising it in another location. This would require further study of the structure of IDH space but by carefully considering quantum mechanical matter progress could be made.

$$F_1 = \frac{M_0 c^2}{h} \tag{7}$$

$$E = \frac{M_0 c^2}{\sqrt{1 - \beta^2}} \qquad \beta = \frac{v}{c} \tag{8, a}$$

$$F_2 = \frac{M_0 c^2}{h\sqrt{1 - \beta^2}} \tag{8}$$

$$F_3 = F_1\sqrt{1 - \beta^2} \tag{9}$$

SECTION 4. THE DIMENSIONALITY OF TACKYONIC SPACE AND ITS HOLOGRAPHIC NATURE

The author suggests the following argument for the dimensionality of tackyonic space. Suppose we take a particle and imaginify its space time description so it will be tackyonic space. Then its velocity will be V_c relative to a stationary observer in 4-space and described by the tackyonic Lorentz transformations (1) and (2). Suppose we let the relative velocity V become zero. Then in this particular case V_c becomes infinite. An infinite velocity has a number of implications. And note that by changing the motion of the observer we could always bring this about. One implication is that the particle will be simultaneously present at *all* points along the imaginary space axis. Thus to specify the co-ordinates of x on imaginification we would need to specify all the imaginary coordinates $(ix_1, ix_2, ix_3, ix_4, ...$ etc). Clearly this would be an infinite set of points (uncountably infinite). Thus the dimensionality of tackyonic space is infinite and I have termed it IDH space (infinite dimensional hyper-space).

Another consequence of the infinite velocity of V_c is that the IDH space has a holographic nature. By holographic information I mean that is has the following two properties.

1. Each part contains information sufficient to reproduce the whole.

2. The magnitude of the part is irrelevant in the sense that each part is of equal status.

Suppose a piece of information is holographic and suppose we split it into two parts A and B. And suppose we send two pieces of information apart with a specific velocity. Consider an observer C who wants to check whether our holographic pieces A and B can reproduce the whole. If he sends a signal with velocity C the speed of light to our observer then that observer could see if the A and B are holographic. But only at the time our observer receives the encoded information. However if the signal traveled with infinite velocity the check could be made instantaneously. The problem with the slow check is that if (as is inevitable) there is some change of A and B in time (eg a quantum change) in their material nature then the observer doesn't notice this. However in tackyonic IDH space when V_c is infinite then because of the nature of such an infinite velocity any matter would be present at all parts of the path of V_c together. We might choose to call this matter which is in tackyonic space information and each location has all the matter present because of the infinite velocity. Thus this information is holographic.

Chapter Two reprinted from the proceedings of the 3rd International Conference Space Time and Gravitation 1994 St Petersbourg Publishing house politechnika,6 Inzhenernaya St Petersbourg 191011 Russia ISBN 5-7325-0381-1

Chapter 3

THE PHYSICAL SIGNIFICANCE OF THE IMAG-NIFICATION TRANSFORMATION, THE PRE-SENTATION OF THE MATRIX ASSOCIATED WITH THE TACKYONIC TRANSFORMATION AND OTHER CONSIDERATIONS

M.H. Sutcliffe
Physics Dept. UMIST P.O. Box 88 Manchester M60 1QD UK

SECTION 1. INTRODUCTION AND PRESENTATION OF THE IMAGINIFICATION MATRIX WHICH SUMMARISES THE TACKYONIC LORENTZ IMAGNIFICATION TRANSFORMATION EQUATIONS

The tackyonic imaginification transformation equations (1) and (2) below were first presented (to the authors knowledge) in Sept 1991 RAS S-Petersburg published in 1994. They were published again in the 1995 proceedings of the 1994 May 22-27 Third International conference on Space, Time and Gravitation (pages 113-116) but without a typing error. The author explained that he thought it was possible that the tackyonic space existed and was not just a mathematical invention. According to the ideas presented here this tackyonic space is infinite dimensional and provides a possible method of using some of the ideas from quantum mechanics. Although it is not possible to derive all of quantum mechanics in this relatively short paper some suggestions are made to enable others hopefully to make progress. This process of transforming to a velocity beyond c is described as "imaginification", the reverse process (described in the May 1994 paper) as "positivisation" because it involved a change of sign from negative to positive of the mass. On imaginification the mass and charge change sign.

The tackyonic imaginification transformation equations (1) and (2).

$$ix^1 = \frac{x - V_c t}{\sqrt{1 - V_c^2/c^2}}, \quad (1A) \qquad it^1 = \frac{t - V_c x/c^2}{\sqrt{1 - V_c^2/c^2}}, \quad V_c = \frac{c^2}{V} \quad (2A)$$

substituting $V_c = c.c/V$ the universal corresponding velocity derived in Sept.

1991 as above these equations become (1) and (2) below.

$$ix^1 = \frac{x - (c^2/V)t}{\sqrt{1 - c^2/V^2}}, \tag{1}$$

$$it^1 = \frac{t - x/V}{\sqrt{1 - c^2/V^2}}. \tag{2}$$

Just as it possible to summarise the ordinary Lorentz transformation equations in a Lorentz transformation tensor matrix **LT** so in the same way it is possible to summarise the equations (1) and (2) above in the imaginification transforming matrix **IT**.

This particular presentation of the Lorentz transforming **LT** matrix uses a mathematical device known as the Minkowski 4-vector. **Minkowski (1909)** where by the matrix **LT** acts upon the so called 4 vector $\vec{a} = (x, y, z, ict)$ and $(\vec{a} \cdot \vec{a} = x^2 + y^2 + z^2 - (ct)^2$. In our summary of equations (1) and (2) we act upon the same 4-vector $\vec{a} = (x, y, z, ict)$ but with the matrix **IT** shown below. The effect of **IT** is to transform the x and t into imaginary ix^1 and it^1 and hence to transform ict, somewhat confusingly into a real quantity $-ct^1$. This new 4-vector we call $\vec{a}imag = (ix^1 iy^1 | z^1 - ct^1)$. Hence the imaginification transformation can be summerised as $\vec{a}$ imag. = IT·$\vec{a}$. Here our 4-vectors are columns.

NB. $(\vec{a}$ imag. $\cdot \vec{a}$ imag.$) = (ct^1)^2 - (x^1)^2 - (y^1)^2 - (z^1)^2$. (The scalar product of $\vec{a}$ imag.

$$LT(\beta) = \begin{bmatrix} \frac{1}{\sqrt{1-\beta^2}} & 0 & 0 & \frac{1\beta}{\sqrt{1-\beta^2}} \\ 0 & 1 & 0 & 0 \\ 0 & 0 & 1 & 0 \\ \frac{-i\beta}{\sqrt{1-\beta^2}} & 0 & 0 & \frac{1}{\sqrt{1-\beta^2}} \end{bmatrix}, IT(\beta) = \begin{bmatrix} \frac{1}{\sqrt{1-\beta^{-2}}} & 0 & 0 & \frac{1/\beta}{\sqrt{1-\beta^{-2}}} \\ 0 & i & 0 & 0 \\ 0 & 0 & i & 0 \\ \frac{-i/\beta}{\sqrt{1-\beta^{-2}}} & 0 & 0 & \frac{1}{\sqrt{1-\beta^{-2}}} \end{bmatrix}$$

where $\beta = V/c$. If we choose to redefine $\beta = c/V$ then **IT**(β) can be given in a simplified form given below.

$$IT(\beta) = \begin{bmatrix} \dfrac{1}{\sqrt{1-\beta^2}} & 0 & 0 & \dfrac{i\beta}{\sqrt{1-\beta^2}} \\[2ex] 0 & i & 0 & 0 \\[2ex] 0 & 0 & i & 0 \\[2ex] \dfrac{-i\beta}{\sqrt{1-\beta^2}} & 0 & 0 & \dfrac{1}{\sqrt{1-\beta^2}} \end{bmatrix}.$$

NB. Det $\mathbf{IT}(\beta) = -1$. Thus $\mathbf{IT}(\beta)$ is an improper orthogonal transformation or reversal.

The inverse of matrix $\mathbf{IT}$ (called in$v$$\mathbf{IT}$) is given by replacing i by $-i$ in the above expressions. The matrix in$v$$\mathbf{IT}$ represents the process of positivisation acting on previously imaginified coordinates. Thus $(\mathbf{IT})(\text{in}v\mathbf{IT})=\mathbf{I}$ is the identity matrix for 4 rows and columns. Thus the three matrices $\mathbf{I}$, $\mathbf{IT}$ and in$v$$\mathbf{IT}$ constitute a three elements mathematical group for any fixed value of beta.

SECTION 2. TACKYONIC SPACE ALSO GIVEN THE NAME IDH SPACE, INFINITE DIMENSIONAL HYPER SPACE

There is a fundamental *difference* between the tackyonic imaginification transformation **and** the ordinary Lorentz transformation which is that the transformation represented by equations (1) and (2) represents a physical change in the state of matter from one kind of matter traveling at velocities less than c to tackyonic matter with an imaginary space-time description and as the author suggested in his two previous papers on this subject with negative signed mass a change of sign of charge and mathematically imaginary energy and momentum which satisfy the equation (3) given below. Tackyons also satisfy equation (4) like light. The ordinary Lorentz transformation does not represent a change of state of matter rather it represents an alternative space-time description of the same state. As such it may be incomplete merely being an intermediate step in the calculation of the tachyonic Lorentz imaginification equations which are given in their simplest form by (1) and (2). Thus the physical significance of the ordinary Lorentz equations may be dubious.

$$(iPc)^2 - (iE)^2 = (M_0^2 c^4), \tag{3}$$

$$E = P \cdot V. \tag{4}$$

where V is greater than c, the velocity of light. I will take as a working assumption that the dual operations of positivisation and imaginification operate throughout space-time at the quantum level of the description of matter. In the 1994 paper I proposed that the frequency with which the process occurs for a given mass M_0 was that given by **L.de Broglie** (1924) for the "internal periodical phenomenon" or IPP. This was given by L.de Broglie to be $F = E/h$ i.e. the rest mass energy divided by Planks constant. As was explained in 1994 by the author L.de Broglie did not say what he thought the IPP might be, but he used it as his starting assumption. From this he suggested that a wave of velocity V_c, the corresponding velocity, as derived by the author without quantum assumptions, would be in phase with the IPP. Also **PAM Dirac** (1932) derived the velocity $V_c = c^2/V$ for the velocity of the state vector of the particle with mass M_0. Few authors have noted that one of the foundations of quantum mechanics is the use of waves or state vectors which travel faster than the velocity of light. Clearly there must be some interactions between the state vectors traveling at tackyonic velocities and the ordinary matter in 4-space because otherwise it would not be possible for these vectors to influence the probability description of matter.

The author takes a different view as to the nature of Quantum Mechanical reality to the conventional so called Copenhagen Interpretation. He proposes that it might be possible to construct something similar to QM using some ideas from tackyonic physics. He suggested in the 1994 paper, that the dimensionally of space beyond the velocity of light was infinite because of the infinite value which arises for $V_c = c^2/V$ when $V = 0$. Because in this case, the particle would be "instantaneously or simultaneously" present at all positions on the ix axis. However it is perfectly possible within an infinite dimensional space such as IDH space for a so called subspace to be defined with say 3 or 4 dimensions. For example the matrix **IT** of the first section may create a 4-dimensional imaginary subspace in IDH space. The author suggests in the last section that our 4-space is in fact a real subspace that is "carved out" from IDH space. One method of creating a finite dimension subspace in mathematics from an infinite dimensional space is to use so called projection operators. The author suggests that the 4 forces act as either projection operators or something similar upon IDH space. I will now show how by using the concepts of tackyonic physics outlined in this paper and the previous two papers a vector space can be derived which has some of the features of the vector space used in QM. A vector space has certain features. It possesses quantities known as "scalars" which can multiply the other quantitity known as "vectors". Given that we have used both real and imaginary numbers in our derivations and results it would seem natural to use the extended complex numbers as out field of scalars, i.e. including infinity + all complex numbers.

The tackor (A) see below

$$
\begin{bmatrix}
-i_\infty \\
\vdots \\
; i_t^1 \\
; i_{i+1}^1 \\
\vdots \\
; i_1^1 \\
\vdots \\
i_\infty
\end{bmatrix} . \qquad \text{(A)}
$$

The vectors which I call tackors are derived as follows. Suppose we have a point which we will simplify to co-ordinates (x^1, y^1) and suppose that we imaginify these c-ordinates in the specific case that $V = 0$. In this case our two frames K and K^1 are coincidental and let us suppose at a specific time say t we apply the transformation described by equation (1). The result of transforming this coordinate is to create in IDH space providing we ignore the (y, z) coordinates the tackor (A) shown above. What it does is it shows all the points along the ix^1 axis in column vector form. However because I have suggested that both converse operations of imaginification and positivisation occur it is necessary to consider what would be result of equation (2) i.e. positivisation on let us say an electron with mass M_0 at $(\dots ix^1, \dots t^1)$. Clearly it would create the real coordinátes (x^1, t^1). Thus from the point of view of 4 space we would only obtain the specific coordinate x^1 at the time t^1. Thus we are justified in rewriting our tackor (A) as the eigentackor (B) shown below. The mathematically conjugate tackor is given by (C) below. This is the equivalent in tackyonic physics of the Schrödinger representation in quantum mechanics.

$$
|\varepsilon> = \begin{bmatrix}
\vdots \\
0 \\
\vdots \\
0 \\
ix_1^1 \\
0 \\
\vdots \\
0 \\
\vdots
\end{bmatrix} , \text{ (B)} \qquad
<\varepsilon| = \begin{bmatrix}
\vdots \\
0 \\
\vdots \\
0 \\
-ix_1^1 \\
0 \\
\vdots \\
0 \\
\vdots
\end{bmatrix} , \text{ (C)}
$$

Definition of the scalar product and of the vector space.

We make the assumption that the field we use is the field of complex numbers with the addition of $\pm\infty$. At a particular time we write our arbitrary tackor as:

$$|\alpha> = \begin{bmatrix} i\alpha_1 \\ i\alpha_2 \\ i\alpha_3 \\ \vdots \end{bmatrix} \quad \text{and write our arbitrary conjugate tackor as} \quad \begin{bmatrix} \overline{i\beta_1} \\ \overline{i\beta_2} \\ \overline{i\beta_3} \\ \vdots \end{bmatrix} = <\beta|.$$

The line above the conjugate tackor entries signifies complex conjugation. The scalar product is defined as

$$<\alpha|\beta> = i\alpha_1 \times \overline{i\beta_1} + i\alpha_2 \times \overline{i\beta_2} + i\alpha_3 + \ldots =$$

$$= \alpha_1\overline{\beta_1} + \alpha_2\overline{\beta_2} + \alpha_3\overline{\beta_3} + \ldots \qquad .$$

This satisfies the following criteria for a vector space

$$1 \ <\alpha|\alpha> \geq 0 \ <\alpha| = 0 \text{ iff } <\alpha|\alpha> = 0,$$

$$2 \ \overline{<\alpha|\beta>} = \overline{<\beta|\alpha>}.$$

We define vector addition in the following and obvious fashion.

$$|\alpha> + |\beta> = \begin{bmatrix} i(\alpha_1 + \beta_1) \\ i(\alpha_2 + \beta_2) \\ i(\alpha_3 + \beta_3) \\ \vdots \end{bmatrix} \quad \text{and scalar multiplication } \lambda \in C \text{ as}$$

$$\lambda|\alpha> = \begin{bmatrix} i\lambda\alpha_1 \\ i\lambda\alpha_2 \\ i\lambda\alpha_3 \\ \vdots \end{bmatrix}.$$

As we are taking the field in our vector space of tackors (which is also in addition an inner product space) to be the field of extended complex numbers or **C**+; it is possible to multiply an eigentackor by an element of fields say $\lambda \in$ **C**+ and obtain another eigentackor which will be a scalar multiple of the first. We make the assumption that all such eigentackors represent the same state in tackyonic space. This means that tackors that specify the same direction in the IDH space I have outlined here form an equivalence class of tackors. This means the actual place or places in the infinite vector that have entries. In other words the actual magnitude of the scalars in the tackor columns is irrelevant because of the properties of complex numbers. We can not say one complex number is greater or smaller than another as I have mentioned elsewhere in this paper and thus the tackors shown below are equivalent.

$$
\begin{bmatrix} 0 \\ 0 \\ 0 \\ 0 \\ ix_1 \\ 0 \\ 0 \\ 0 \\ 0 \end{bmatrix} \text{1st place} \quad \sim \quad \begin{bmatrix} 0 \\ 0 \\ 0 \\ 0 \\ ix_1\lambda \\ 0 \\ 0 \\ 0 \\ 0 \end{bmatrix} \text{1st place.}
$$

Similarly

$$
\begin{bmatrix} \vdots \\ 0 \\ ix_1 \\ 0 \\ 0 \\ 0 \\ ix_2 \\ 0 \\ \vdots \end{bmatrix} \begin{matrix} \\ \\ \text{1st place} \\ \\ \sim \\ \\ \text{4th place} \\ \\ \end{matrix} \begin{bmatrix} \vdots \\ 0 \\ ix_1\alpha \\ 0 \\ 0 \\ 0 \\ ix_2\beta \\ 0 \\ \vdots \end{bmatrix} \begin{matrix} \\ \\ \text{1st place} \\ \\ \\ \\ \text{4th place} \\ \\ \end{matrix}
$$

where $\alpha, \beta \in \mathbf{C}+$ are equivalent.

SECTION 3. TACKYONIC SPACE. I D H SPACE. ITS NATURE

I have outlined one way of beginning to develop the mathematics of the space beyond the velocity of light. I will make some further comments as to the nature of the space. I said in the 1994 paper that matter traveling beyond the velocity of light would have a holographic nature. This is because of the property of an infinite velocity. If our expression for velocity beyond the velocity of light is correct than from $V_c = c^2/V$ when $V = 0$ there is effectively an infinite velocity. The implication of this is that there is no "before" and "after" in IDH space because any matter would be present at all points along its path. This matter which we might choose to call tackyonic information would thus be present at <u>any</u> and <u>all</u> points and thus we would be able to describe it as holographic because each point would have the same status and each would have effectively <u>all</u> the information of the other point.

Tackyonic space is more general than a Hilbert space. IDH space is not a Hilbert space because it need not have a finite inner product. This is because we allow infinite numbers in our field scalars which may be strictly mathematically rigorous. The reason we have to allow these infinite numbers is because of the infinite velocities that arise which might also give rise to infinite imaginary displacements. However the IDH space I have developed has many of the features of a Hilbert space and we can consider it in that way most of the time.

IDH space helps to answer an important question that is sometimes asked by students of QM. For example **C.Antonopoulos** in the same 1994 RAS publication [1] asks about the problem of the particle which makes a quantum change from E_1 to E_2, i.e. energy change. He asks about the problem of the time it takes to make the change. But it could also be asked where does say the electron go to? Where is the electron when it changes energy states within the orbit of a nucleus. QM traditionalists don't answer this question because they claim it shouldn't be asked or its not answerable because it contradicts the uncertainty relations. My answer in this version of new QM or tackyonic QM is simple. The electron **enters into IDH** space. It **may** be replaced by a different electron with a photon at a different place. This is what a quantum change is. The time it takes to do this will be in keeping with the frequency of the IPP of L.de Broglie mentioned earlier.

Tackyonic space also helps to explain the so-called EPR paradox about the interconnection between two apparently disconnected atoms that had previously been part of a diatomic molecule. It suggests that the apparent interconnection that occurs in 4-space occurs because of the holographic nature of IDH space that is outlined above.

In this paper I have not discussed the role of linear operators acting on tackors. Linear operators in QM often have a factor of $-ih$ or ih in front of them. The imaginary factor has the effect of crossing the light velocity and this if acting on an eigentackor it has the effect of positivising one coordinate. The significance of h Planks constant has been partly given by L. de Broglie above. It would seem reasonable that if we put all points of space-time on the same footing they would have the same constant of proportionality associated with them. However the author does not fully understand the exact manner in which this transformation occurs. If rather than using the Schrödinger i.e. displacement tackor method we were to use the momentum energy 4 vector we could create a momentum representation instead.

Another feature of QM is so-called unitary transformations. The transformation **IT** from section 1 is not a unitary transformation because it has determinant -1. Thus it represents a so called reversal rather than a rotation. It is well known that a unitary transformation in a complex vector space is one that leaves invariant the sum of the squares of the modulii of the variables, i.e. the product $(\overline{X}_i \cdot \overline{X}_i)$ where $\overline{X}_i$ is the complex conjugate of the column vector X_i. The question that arises is what is the physical meaning of a similar such transformation in IDH space. I suggest it would leave invariant a quantity representing the infinite dimensional equivalent of the length squared of the path of light in IDH space.

The nature of light itself in IDH space is another area of interest. It is well known that in addition to the real **E** and **B** fields which satisfy Maxwells equations there are another imaginary and complex solutions to Maxwells equations. I suggest that this other mathematically imaginary light exists in IDH space. It has imaginary energy iE and imaginary or complex $\mathbf{E}$ and $\mathbf{B}$ fields. The transformation equations are given by replacing V by V_c and E and B by iE and iB in the components of these vectors in the Poincare or Einstein formulae. Thus there are at least two kinds of light only one of which exists in 4-space.

SECTION 4. THE SPACE-TIME ISOMORPHISM CONJECTURE

At present there are 4 forces known in nature: the Short nuclear, Long nuclear, Electromagnetic and Gravitational forces. The conjecture of the author (which would be proved wrong immediately a fifth independent physical forces were discovered) is that there is a correspondence between the four forces and the 4 dimensions. He suggests that gravity would correspond to time i.e. t and that the other three forces would correspond to the three space dimensions i.e. (x, y, z). One suggestion as to how this happens **is that the 4-forces mediate and create the space-time structure from the IDH space mentioned, long nuclear and electromagnetic forces mediating the 3 space dimensions.** At present gravity is a uni-directional force only being able to attract, similarly time flows in one direction only. This justifies the association of time with gravity because they are both un-directional in a sense. (It has been claimed that velocities greater than c are impossible because of this quality of time. However this is based on a misconception about the nature of imaginary time. Imaginary time does not have a "before" and an "after" because of the nature of imaginary numbers. No imaginary number is either greater or less than another. Only the magnitude can be ordered and has this quality but the magnitude is not an imaginary number.)By contrast the electromagnetic force is able to both attract and repel as the short and long nuclear force must be able to because otherwise the nuclear would collapse. Thus we feel justified in relating these forces to the three space dimensions. This idea is tentative but provides a contract to the existing orthodoxy in cosmology. At present the existing orthodoxy in cosmology is that space-time "began" with the so-called big bang and the 4 forces separated with the first few micro-seconds into 4 from an original single unity or singularity. The BBM big band model replaced a theory created by Hoyle which was the so-called continuous creation of matter model or CCMM. What this conjecture is suggesting is that irrespective of whether the BBM or CCMM is correct or some other model the space-time structure us being mediated today by the 4 forces which act on IDH space. It has been claimed that because clocks are altered by varying gravitational field this proves GRT. It actually shows there is a relation between time and gravity which is also suggested by this conjecture.

REFERENCES

1. **Antonopoulos C.** Time; The Problem concept of the Quantum Theory. Published RAS ST Petersburg, 1995, p.162.

2. **Louis De Broglie**. Phil. Mah., 47, 1924, p.446.

3. **P.A.M. Dirac**. The principles of Quantum Mechanics. Oxford University Press, 4th Edition, 1947.

4. **Minkowski H.** Raum und Zeit (Space and Time) lecture Cologne, Congress of Scientists, 21 Sept. 1908, published Phys. Z., vol.10, 1909, p.104.

NB In the Appendix some further derivations obtained from the unpublished paper upon which this paper is based along with CHAPTERS 4 and 5 will be set out.

Chapter 3 from POLITECHNIKA St Petersbourg RAS Sept 16-21 1996 Selected papers 4th International Conference Problems of Space Time and Gravitation. ISBN 5-7325-0450-0 published 1997 Russian Academy of Sciences

Chapter 4

THE "INTERNAL PERIODICAL PHENOMENON" OF LOUIS DE BROGLIE; IS IT THE TRANS-FORMATION OF MATTER FROM MINKOWSKI 4 SPACE TO IDH SPACE? AN UNPUBLISHED PAPER FROM MAY 1991

M.H. Sutcliffe
Physics Dept. UMIST P.O. Box 88 Manchester M60 1QD UK

INTRODUCTION

In the keystone paper by Louis de Broglie (1924) an important assumption is made that an "internal periodical phenomenon" occurs for all matter "particles". No explanation is given for the nature of the "internal periodical phenomenon". However, the clear assumption of the paper is that the "internal periodical phenomenon" is a universal phenomenon with a frequency that L. de Broglie assumes to be related to the rest mass of the particle by the formulae $moC^2 = h\mu_0$ where μ_0 is the frequency of the "internal periodical phenomenon" taken in a frame at rest relative to the rest mass mo (C being the velocity of light and h planks constant).

The author will re-examine these assumptions and make a fresh suggestion as to the nature of the "internal periodical phenomena" namely the affirmative to the question posed in the title.

SECTION 1: THE RESULTS OF LOUIS DE BROGLIE.

Louise de Broglie started from the assumption that all the energy moC^2 would be ascribed to an "internal periodical phenomena", (this was on the basis of an analogy with light each photon of which has energy given by hf where h is planks constant and f is the frequency of the relevant photon). Clearly, since all particles traveling at velocities less than C have mass, this "internal periodical phenomena" must be universal for all have mass in the cosmos.

L. de Broglie than stated that if we consider the mass mo (in its rest mass frame) to be moving relative to a stationery observer then the observer would observe the frequency as

$$\mu_0(\sqrt{1 - V^2/C^2}).$$

An obvious result from special relativity. (Any problem that might arise could only come from the quantum nature of the observation at the "particle" level. The use of the word "particle" suggests the mass would be effectively point-like and to actually measure such a small time interval might present experimental difficulties and perhaps give rise to uncertainties. However, we will ignore any problems here noting that when L. de Broglie made this important discovery, Heisenberg's uncertainty principle had not been discovered.)

L. de Broglie then suggested that a wave traveling with phase velocity C^2/V where V would be the velocity of the mass relative to the stationary observer would be in phase with the "internal periodical phenomenon". It would have wavelength h/p where p would be the momentum and frequency given by $\mu = E/h$. The energy E for the mass mo being considered by a stationary observer, departing with velocity V from that observer; similarly for p.

The group velocity of the wave was found to be V. In the paper L. de Broglie said that the wave could not carry energy according to Einstein's ideas. However, in his derivation, as we can see, he made quantum assumptions to obtain **his** derivation. This would seem to suggest that at the most fundamental level special relativity and quantum theory are inter-related. The author has been able to derive the velocity in a simple and direct fashion *without* using quantum assumptions. The author sees the velocity as being of fundamental importance and calls it the "corresponding velocity".

SECTION 2: AN EXPLANATION OF THE "INTERNAL PERIODICAL PHENOMENON".

Having derived the corresponding velocity C^2/V in a manner that only depends on the assumption that the co-ordinate of matter imaginify the author will proceed to put a bit more definition on the concept of imaginification. In the paper " the special theory of positivity", the author suggested that there is a universal on-going interaction between IDH space and 4-space and he assumed that this took place through an exchange phenomenon. The **assumption that the interaction is a one for one exchange may not be necessary to account** for the experimental facts and is only one possible interaction.

What is important in these ideas is the assumption that IDH space and 4-space are in **continuous interaction** occurs. In other words, the author believes this "internal periodical phenomenon" to be an existing physical process that is universal. It is the process of matter at the quantum level interacting with the IDH space. when the matter at the quantum ie. point-like particle-type level has "exchanged itself" into IDH space, it will have velocity C^2/V as the author and L.de Broglie agree. The frequency of this interaction is given by L. de Broglie as above. (The special theory of positivity ;,an unpublished paper from 1988 mentioned above.)

Chapter 5

THE ANALOGY OF THE LORENTZ TRANS-FORMATION IN I.D.H. SPACE; IS IT THE UNITARY TRANSFORMATION? AN UNPUBLISHED PAPER FROM MAY 1991

M.H. Sutcliffe
Physics Dept. UMIST P.O. Box 88 Manchester M60 1QD UK

It is well-known that a Lorentz transformation is an orthogonal transformation. If x, y, z are cartesian co-ordinates and t is the time of the event in a space-time diagram, then a Lorentz transformation acts in such a way as to keep $x^2+y^2+z^2 = (ct)^2$ invariant in changing from one frame of reference to another. This is the so called constancy of the velocity of light, one of the two cornerstones of the special theory of relativity.

If we adopt the well-known convention $x = x_1, y = x_2, z = x_3$ and $ict = x_4$ then the effect of a Lorentz transformation which is a group that can be represented by a matrix is such as to keep x_i^2 constant (Einstein summation convention in operation) in transforming from one frame of reference or co-ordinate system in Minkowski 4 space to another. Such transformations that keep the sum of the squares of the co-ordinates of a point in a co-ordinate system invariant under transformation are known as orthogonal transformations. Conversely if the transformation has this property, then it is orthogonal.

However, the question that suggests itself is what transformation would be analogous to the Lorentz transformation in IDH space?

One feature we would look for if we wished to discover an analogy physically would be some kind of 'world distance' that would be kept invariant. We would need physically some concept of frame of reference with respect to which this new 'world distance' would be kept invariant. Because IDH is infinite dimensional the transformation (at maximum generality) would have to be able to act in infinite dimensions.

We would expect this transformation to be a group because if we were to combine one such transformation with another mathematically, this would be equivalent to two successive changes of frame of reference in IDH space. We would expect these two successive changes to be equivalent to one single change of reference that combined the two. Because of this physical interpretation we would expect the change to be reversible and this would correspond to the inverse element within a group. We would at the first stage assume the transformation to be linear because of the linear nature of the Lorentz transformation.

It is well-known result from the theory of the group of representations of linear transformation that unitary transformations have the following property: they leave invariant the sum of the squares of the moduli of the variables, ie. they leave invariant the product $X^T X$ where X is the complex conjugate matrix of X with 1 column and n rows. If we write U for the unitary matrix which represents the group element Ug say, then this leads to the well-known property as is stated by P.A.M. *Dirac* that $U^T U = 1$.

The matrix is in general able to act infinite dimensionally. The analogue for the world distance would appear to be the sum of the squares of the moduli of the coordinates of a point in IDH space. Under a unitary transformation this summation would be invariant. Thus a unitary transformation would correspond to a change in the frame of reference. Such a change could be a rotation in IDH space or a translation, or even perhaps a reflection. The difference with the Lorentz transformation is this: the Lorentz transformation except in the trivial case has an explicit velocity dependence, however, it would be possible to have non-trivial translations of frames of references that didn't depend on any velocity. That is because different points on a line have difference eigenvectors associated with them in IDH space. If we consider the relative velocity to be zero between them in 4-space, then in IDH their velocity is infinite by the well-known relation C^2/V when $V \to 0$. Thus the eigenvectors themselves are engaging in continuous ongoing non-trivial translations, and unitary transformations are connecting them from one point to the next in IDH space.

Note added 2010 November

This suggestion for the physical meaning of the unitary transformation ie. that one point in IDH space is transformed to another with the infinite sum of the squares of the variables being kept constant, is not dependent on S.R.T. This suggestion is independent of the correctness of the Lorentz transformation.

The "well known result" can be found THE THEORY OF SPINORS ÉLIE CARTAN DOVER PUBLICATIONS, NEW YORK 1981 ISBN 0-486-64070-1.

Chapter 6

THE DUAL NATURE OF LIGHT

Let us examine the nature of matter ie photons etc that travel at the speed of light.

We will only consider matter that satisfies Maxwell's Equations.

A PLANE POLARISED PROGRESSIVE WAVES

It is well known that the real part of the waves

$$\mathbf{E} = (\mathbf{A}(e^{ik\mathbf{e}.\mathbf{r}}))e^{-iwt}$$

$$\mathbf{H} = \mathbf{e} \wedge (\mathbf{A}(e^{ik\mathbf{e}.\mathbf{r}}))e^{-iwt} \qquad e^{i\theta} = \cos\theta + \mathrm{i}\sin\theta$$

$\mathbf{e}$ is a unit vector, assumed real

t is time

$i^2 = -1$

k = wave number = $\frac{2\pi}{\lambda}$ $c = f\lambda$.

$\mathbf{r}$ = polar distance from origin

$\mathbf{A}$ = amplitude vector

w = angular frequency = $2\pi f$

satisfies Maxwell's equations.

However it should be considered that the imaginary part also satisfies Maxwell's equations. It travels with the same velocity and has a phase difference of $\pi/2$ radius. As it is of mathematically imaginary nature we will not perceive it. The dimensionality of IDH space in which the imaginary components exists in full generality is infinite. This is demonstrated in Chapter Three.

ie The following component.

$$-\mathbf{A}i \left(\sin k\mathbf{e.r} \times \sin wt \right) = \mathbf{E}$$

$$-\mathbf{e} \wedge \mathbf{A}i \left(\sin k\mathbf{e.r} \times \sin wt \right) = \mathbf{H}$$

exists in I.D.H. space.

B PLANE EVANESCENT WAVES

It is also well known that we can obtain more general wave selections than the plane polarised progressive waves that still satisfy Maxwell's Equations. This is done by assuming that $\mathbf{e}$ and $\mathbf{A}$ are complex. Such solutions contain a mixture of real and imaginary components. We make the assumption that both the real and the imaginary solutions have actual physical significance. Both exist and are complementary aspects of the light waves. It seems reasonable to assume that photons have a dual nature ie complex nature.

Such an example of a wave is

$$\mathbf{E} = (0,0,1)e^{ikch\beta x}e^{-ksh\beta y}e^{-iwt}$$

$$\mathbf{H} = (ish\beta, -ch\beta, 0)e^{ikch\beta x}e^{-ksh\beta y}e^{-iwt}$$

$$\mathbf{A} = (0,0,1)$$

where $\beta > 0$ real $i^2 = -1$ other functions as previous in (A).

Chapter 7

A NEW COSMOLOGY FOR OUR TIME

From the time of the ancients up until the present day there has always been a debate as to the nature of the Cosmos. One key debate is the simple question 'is space empty or is it full?'

Parmenides suggested that space was in some sense full and consisted of the so called plenum out of which reality emerged. Others argued that space was a vacuum it was largely empty except for matter made up of atoms "Democratus". There was very little in the cosmos other than small fragments of matter separated by the eons of empty space. After atoms were identified by DALTON and MENDELEEV, space became even more empty because even the atom itself became largely empty space, the volume of the nucleus being tiny in comparison with that of the atom and the electrons being treated as point like negatively charged particles. On further investigation even the nucleus of an atom contains empty space, the protons and neutrons do not account for all the volume of the nucleus. So much empty space in the cosmos, it is a surprise anything exists at all!

In addition to this we are told that the universe begun from a so called Big Bang, where all the matter was concentrated into a singularity and then expanded out from it. It makes no wonder that matter takes up virtually no space today because it took virtually no space at all up at the origins of the universe, according to the Big Bang theory.

The key questions that those who advocate that the universe is full have to answer are simple. Firstly where is all the matter or material with which the universe is filled?, and secondly what is its nature? How can we observe it if that is possible?

Clearly these questions are not easily answerable because no definitive answers have been given since the time of Parmenides. In a tentative way these questions are the ones this book attempts to answer.

The author asserts that the vast part of the cosmos is hidden. All we observe is the seen part of the far greater reality most of which is unseen.

7.1. The Tachyonic nature of the unseen reality

The obvious question is if there is a vast underlying hidden aspect of reality why can we not see it or perceive it. The author suggests that this is because its nature is tachyonic, it travels too fast for us to be able to see it.

It is an underlying substance of immeasurable mass due to its vastness. It connects everything together. Whether the matter is living or dead, all is connected by this wonderful substance. The energy travels beyond the speed of light and hence we humans cannot observe it, with our eyes but that does not mean it does not exist. It simply means we have not yet understood its full nature. As we slowly discover this wonderful substance and energy; we will be able to move across the galaxies. This is unlikely to happen in the 21st century but equally it can not be entirely dismissed. As is explained in CHAPTER 3 this unseen energy also helps to explain the so called holographic universe an idea that this author first came across, in the work of Professor DAVID BOHM.

The holographic nature of the universe is discussed in one of his finest works entitled "Wholeness and the Implicate Order" published Ark Paper Backs, Ark Edition 1985 Routledge and KEGAN Paul, London WC2H, ISBN 0-7448-0000-5. In this particular book BOHM insisted that nothing could travel faster than light but also strongly insisted that there was an interesting nature to reality. This so called underlying reality he called the implicate order. Perhaps D. BOHM suspected its nature to be tachyonic but at the time of first publication he was working within the scientific orthodoxy of the time, when few were prepared to question the then widely taught theory that this author has questioned since 1982/1983 namely the theory that nothing can travel faster than light.

The weakness of DAVID BOHMs argument set out in page 122 which this author will briefly mention namely that if matter could travel faster than light ie. for example an atom then the atom would fall apart because the atom is held together by electromagnetic coulombic force $F = C\frac{Q_1 Q_2}{r_2}$ is the assumption that real matter could not be imaginified in the language of this author, travel faster then light to another part of the cosmos and then be positivised.

Another point BOHM makes is that the speed of light is effectively, a signal. "The speed of light is taken not as a possible speed of an object, but rather is the maximum speed of propagation of a signal".

However BOHM then by his use of the word 'perhaps' in the following quote page 123 indicates his own doubts as to the argument he has provided. "A signal is indeed a kind of communication. So in a certain way, significance meaning and communication become relevant in the expression of the general descriptive order of physics (as did also information, which is, however, only a part of the content or meaning of a communication). The full implications of this have *perhaps* not yet been realized. ie. of how certain very subtle notions of order going far beyond those of classical mechanics have *tacitly.* been brought into the general descriptive framework of physics".

The purpose of this approach is to help the reader to understand a little more what BOHM is hinting at with his use of the word "perhaps" and "tacitly". What this author is emphasizing is

1. information can be transmitted faster than light, there is nothing in the definition that says it can't be as is pointed out in CHAPTER 1. ie. the definition of information does not preclude it from being transmitted beyond C, the velocity

of light.

2. that this 'information' makes tachyonic physics an important area for study.

Other authors have also discussed the holographic nature of reality. For example WILBER New Scientific Library. *The Holographic paradigm and other Paradoxes. Exploring the Leading Edge of Science.* Edited by KEN WILBER ISBN 0-8777-235-3. The special understanding of this author is explained in CHAPTER THREE, concerning the so called Holographic nature of reality. This author emphasizes that in order to have a holographic universe or a holographic nature to the universe, the connection must occur at tachyonic velocities. This does not discount that the hologram created by GABOR on this side of the light barrier represents a key discovery in so called real matter. Because it was GABOR'S discovery of the hologram that lead authors like D. BOHM and K. WILBER to write about the holographic nature of the universe, or rather to use this concept of a holographic paradigm. D. BOHM used the phrase implicate order. **This author's simple suggestion is simply that the implicate order which is holographic in nature out of which D. BOHM said real matter emerged is tachyonic.**

7.2 Some Considerations on the Nature of Light.

In CHAPETR 6 this author observes that imaginary solutions also satisfy the Maxwell equation. This was first noted by the author when listening to lectures in Electromagnetic Theory and Special Relativity by Professor GREGORY at Manchester University, who pointed out that these discarded imaginary solutions also travel at the velocity of light within the Maxwell Equations. ie Professor GREGORY showed some of these solutions to the author.

The fundamental proposal of the author is that light can be dual nature. It is both real ie observable and imaginary ie unobservable. Thus these are effectively three types of light. Let me explain. There is purely real light. There is purely imaginary light and there is combined ie real and imaginary light. ie. L, IL and $L + IL$.

Thus light serves both to show, the real matter we observe ie. in its real form and also the imaginary form acts as a lower velocity limit to tachyons. Thus in some sense the nature of light is from the purely physics point of view both used to observe real matter and also acts as a lower velocity limit to that information that travels faster than the speed of light.

The issue of the constancy of the speed of light or otherwise would not alter the question as to whether it was possible to travel faster than light.

The author sees this book as only a very humble beginning on the path to a more comprehensive understanding of tachyonic physics. The derivation given in CHAPTER 1 of the equation $V_C = \frac{C^2}{V}$ that clearly shows a link between tachyonic physics and quantum mechanics starts from the Lorentz transformation equations. In CHAPTER 3 it is suggested that the Lorentz transformation equations are useful as an *intermediate* step to derive the tachyonic transformation equations. Thus it is not entirely necessary to assume the constancy of the speed of light. It is a rough guide to say it is constant and then from this the $V_C = \frac{C^2}{V}$ formulae can be obtained.

As to the particular value of C that is used this will actually be determined by the particular part of the cosmos that it relates to. It would be making a massive assumption to assume that the speed of light is constant everywhere in the cosmos. In fact there is virtually no evidence to back this up nor would there be until such time as such experiments could be carried out in far away parts of the cosmos. It would be unscientific nonsense to believe that the speed of light would be constant everywhere in the Universe. Because one of the basic principles of science is that results are experimentally testable. How could we possibly test that assumption ie. the assumption that the speed of light is constant in a vacuum in another part of the cosmos unless man had fully mastered the practical and theoretic aspects of tachyonic physics. And as this author has mentioned this cosmology is only a humble beginning on the road to that goal!

But simply to recognise the interconnected nature of the cosmos as well as its immerse complexity is something humanity needs to appreciate more. The comprehensive ability to tachyport (ie travel faster than light) across the cosmos may well take centuries to discover. The cosmos is interconnected and in the humble opinion of this author this occurs partly in the tachyonic sphere.

Clearly the so called butterfly effect whereby according to atmospheric physics research a small motion in one part of the earth's atmosphere results in change in the earth's atmosphere in another, demonstrates the interconnected nature of the climate system achieved at slower velocities. (ie less than the speed of light.) However this author strongly suggests that the interconnected nature of the vastness of the cosmos occurs through tachyonic means. Also biologists, life scientists generally are well as climatologists are well aware of the interconnected nature of earth based matter some in living systems offering more physical or geological examples. An earthquake in the pacific can create a Tsunami that takes hours to reach the shores of India or the connecting oceans. But these connections from the original event that created the Tsunami (ie. Linking the earthquake to the shores of India) are vastly slower than the speeds that would be necessary to, for example reach the nearest star ie Alpha Centurion approx 4 light years away.

Thus if humanity were ever to travel to even the nearest star he/she would have to master the ability to tachyport ie. transport faster than the speed of light. Clearly this would be a huge task that could take centuries to develop but is the opinion of the author it is at least theoretically possible. But the significance of this work lies not in proposing the theoretical possibility of reaching distant stars rather in understanding an alternative way at looking at the existing equations of physics.

7.3 Discussion of the relationship between tachyonic physics and quantum mechanics

In some sense what this book does is to attempt to bring out the meaning of the already existing equations within quantum mechanics. There is common ground between the two approaches. In fact some experiments already performed at the quantum mechanical level *if correctly interpreted* already demonstrate the ability of signals to move faster than light and some show small particles have been transported faster than light so called relocation experiments have been carried out. These specific electron relocation experiments will be discussed, in a **Third** edition of this book, but the author interprets them as demonstrating tachyonic matter transmission.

Some other experiments will be discussed later in the book. The interpretation the author is putting on these experiments is controversial but since the whole book is likely to be controversial if not anathema to many career and orthodox scientists, then this will have to be lived with. It is in the nature of the new thinking that it is frequently not accepted at least initially. But every orthodoxy should be capable of defending itself against mistaken new thinking by showing the obvious flaws in such new thinking. What was shown in CHAPTER ONE was the errors that occurred in the so called proof that matter can not travel faster than light. From the experience of the author such mathematical proofs saying that matter cannot behave in certain ways have to be treated warily. For example in the quoted book by BOHM (wholeness and the implicate order) BOHM examines the proof by Von Neumann against hidden variables in Quantum Mechanics. What BOHM shows is that the Von Neumann is flawed because of the assumptions that go into the proof. Of course BOHM's disproof does not of itself mean there are hidden variables in quantum mechanics merely that there can be and they are not ruled out mathematically.

As has been mentioned in CHAPTER ONE and elsewhere the C^2/V formulae where $V < C$ which is derived without quantum assumptions in CHAPTER ONE has also been derived with quantum assumptions by P.A.M. DIRAC independently and earlier by Louis de Broglie. The meaning Louis de Broglie gives to C^2/V is somewhat different to that P.A.M. DIRAC gives. As I have mentioned in CHAPTER 3 and 4 Louis de Broglie starts from what is **effectively** an arbitary assumption about so called internal periodical phenomena without telling us what this I.P.P. is. My suggestion is that the I.P.P. consists of an exchange of matter into the tachyonic space and back occurring at the quantum level. Thus the I.P.P. links quantum phenomena with tachyonic space, according to this approach.

When we look at the derivation given by P.A.M. DIRAC where we next find the derivation of our C^2/V formulae we find here that C^2/V is the velocity of the state vector. Other author such as BOHM tell us that in some senses the state vector represents a description of the experimental set up. It represents the different possible outcomes from which a specific experimental outcome is extracted. How could such a theory be meaningful at all unless the state vector traveling with a tachyonic velocity was able to interact with real slow speed matter as we perceive it? Thus this author suggests that tachyonic explanations are built into the structure of quantum mechanical theory, a theory that has had many experimental successes. When a linear operator acts on the state vector to extract the real eigenvalue it is acting on a tachyonic 'state vector description'. In the language of this author the linear operator positiveness the real matter from its imaginified state.

When an electron or other subatomic particle imaginifies at the quantum level it effectively vanishes from our $3 + 1$ dimensional slow speed real matter world.

The author argues in his chapter 3 paper that this is what happens when an electron changes from one energy level to another within the atom. In the language used by David BOHM in his book "Wholeness and the implicate order" when the electron imaginifies its space-time description it becomes part of the *implicate order* and in the language of this author it becomes tachyonic. When the electron reappears (after having vanished into the implicate order) it becomes once again part of the explicate or manifest world. BOHM pointed out that the root of the word "manifest" is 'at hand', thus the manifest or tangible world is the one we can grasp because it is at hand. This process of making an electron explicit, manifest or tangible the author described as the term positivisation. This is because the author suggest from looking at the equations that the mass becomes negative in imaginifying into tachyonic space and thus when it returns its sign will become positive.

The negative sign of the tachyonic matter can also be interpreted as such matter being taken away or vanished from our manifest explicate world. It is also suggested that charged particles change in the process of imaginifying into the tachyonic space, by switching signs.

On returning from tachyonic space into the manifest $3 + 1$ space the charged particle changes its sign back again.

Thus the *internal periodical phenomenon* is constantly linking the tachyonic world and $3 + 1$ space at the quantum level. Whether there is always a one to one correspondence between the departing and arriving particles (ie to and from tachyonic space) is not a question the author is prepared to give a definitive answer to. Clearly most of the time there must be otherwise there would be a violation of mass conservation occuring regularly and observably. But if HOYLES CCM model were ever to become widely accepted (continuous creation of matter model) then the hypothesis of this author would be that matter was 'leaking through' from tachyonic space more being left than returning to tachyonic space.

7.4 Discussion of the role of tachyonic space in providing a matrix for the storage of information and discussion of relevant ideas of Dr Rupert Sheldrake.

The full exposition of these ideas can be found in the recently republished book "A New Science of Life" Author R. SHELDRAKE.

Briefly R.SHELDRAKE proposes a concept of morphic resonance and also morphogenic fields. The morphogenic field informs the matter how to form or shape itself. The root word morph is commonly used as a verb to imply "taking on the shape of" R Sheldrake could not understand how biologically sophisticated creatures could grow from an amorphous collections of cells, amorphous meaning without form or of low complexity of form. Although many of his examples come from biology he also drew examples from chemistry.

The common explanation is that cells form themselves into their fully formed or adult shape by following the instruction code contained within the DNA of each individual cell. The question a critic of the DNA theory might pose is: given that each cell contains the same instruction code (ie the same DNA which is replicated on cell division) why should some cells become eye cells in one place ie the head while others become skin cells on the bottom of the feet? The whole complex structure and arrangement of the cells to grow let us say a frog is supposedly contained within each cell. Not only is this the case but the whole process of differentiation of the cells from the embryo, and the various stages of development of the embryo the whole process, the roles of the cells, their differentiation into distinctive organs, this whole amazing process is supposedly *entirely* controlled and informed by the DNA.

This is what R. SHELDRAKE tried to explain, so called morphogenisis, the creation of form.

R. SHELDRAKE'S explanation for the cause or creation of form ie shape, the detailed arrangement and structure of matter within both living and none living matter was to propose that all matter was governed by so called morphogenic fields. These morphogenic fields were effectively a record of previous matter forms that then influence the present form as it was created. The more the previous form had occurred the stronger its influence. As a simple example if ten blue new crystals had grown and three red ones, then the fourteen crystals was 3/13 likely to be red and 10/13 to be blue.

Morphogenic fields operate according to R. SHELDRAKE at all levels from the formation of a galaxy (although he doesn't go that large) right down to the fields within an atom or a nucleus. The way that a galaxy develops is determined by previous galaxy formations, just as the way that an embyonic frog develops is determined by previous same species frogs have developed.

Additionally R. SHELDRAKE proposes that his morphogenic fields are transmitted across space so a field on one side of the world can influence matter formation on another side of the world, as can past events.

The clear question that needs to be posed about R. SHELDRAKE M-fields is where are they? or where are the stored? If they exist (and R. SHELDRAKE quotes evidence suggest they do) although they are only a hypothesis it would seem reasonable to enquire as to their nature.

A reader could probably guess from the subject matter of this book what the author suggests is the answer to this question. Yes, you've guessed it, the author suggests that R. SHELDRAKE'S M-fields or morphogenic fields are stored in tachyonic space. This would explain how they are able to instantaneously transmit to where they are required. Thus tachyonic space which is also described in the 1996 RAS paper CHAPTER 3 as I.D.H. space acts as the space in which the M-fields are stored, effectively the M-fields are an information matrix existing in I.D.H. space. Another way of explaining this is to say that I.D.H. space is the ground which acts as the storage site for the information matrix of R. SHELDRAKE'S M-fields.

Thus this author has provided a viable explanation as to why Rupert Sheldrakes ideas have a valid physical explanation.

Rupert Sheldrake's work is seen as fundamental by this author is fundamental to the new cosmology that is emerging in the 21st Century. This is the cosmology that links the smallest to the largest and the far away galaxies to our planetary system. The nature of the cosmos is holographic, it is interconnected and new discoveries about its nature are ongoing.

REFERENCES

R. Sheldrake, A New Science of Life, ICON BOOKS LTS 2009, ISBN-10-1848310420

Chapter 8

DISCUSSION OF BELL'S THEOREMS

1 Introduction

Bell's theorem is confirmative of tachyonic physics, according to this author but clearly a reader would wish to have further discussions than this somewhat bold statement.

John S Bell (deceased 2004) was particularly interested in the so called EPR paradox (Einstein Podolsky Rosen) reference at the end of this chapter. In its simplest form the EPR paradox concerns a diatomic molecule. Suppose one atom was to split and fly off in one direction then the other would fly off in the opposite direction. This would happen due to Newton's Second Law. The cause of the split in the diatomic atom is irrelevant to the discussion. But the main effect would be for one atom to move in one direction and one to move in the opposite. (Clearly with modern scanning tunnel microscopes which the author first saw in Russia in the 1990's this is not just theoretical: atoms can be observed or to be more accurate we should say a computer generated image of an atom can be observed.)

The observation of the departing atom would be different from different angles as is illustrated below in FIGURE 1.

In the language of physics this is explained as different frames of references will show the twin atoms departing in different ways. A somewhat obvious statement. But there are two observation points where it would not be possible to observe the two atoms departing in a purely theoretical and idealised model.

These are when our notional observer is collinear with our atoms A & B.

However since physics both theoretical and practical allows room for variation ie of experiment and this collinear observation is only made in passing.

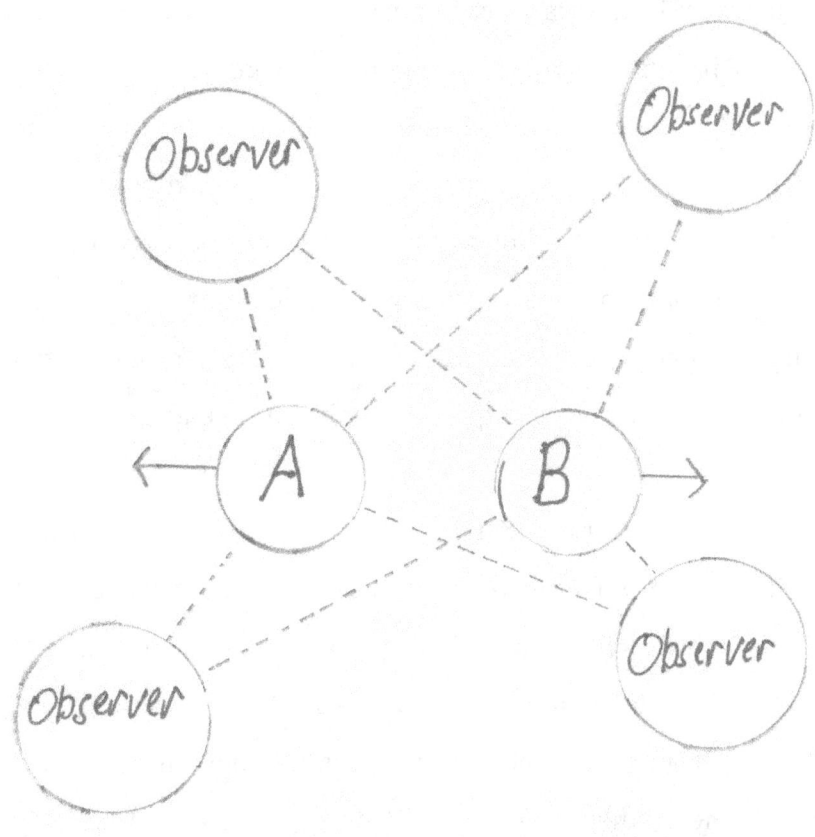

FIGURE 1. SECTION 1.

47

All so far seems straightforward to the reader the author would hope. So where is the mysterious paradox? The answer lies beyond Newton in the world of quantum mechanics. However Newton still plays a role in this author's explanation of the EPR paradox which was used by the authors of the EPR paper to shine a light on weaknesses within Q.M. [All theoretical models have a degree of incompleteness.] The actual method we are using to explain the EPR paper is partially attributable to D.BOHM.

Atom A would have a momentum call it p then because momentum is a vector quantity and atom B is the same mass going in the opposite direction then atom B must have momentum $-p$. Thus according to the explanation presented in the EPR paper when our observers knows the momentum of atom A the observer also knows the momentum of atom B.

Suppose equally an observation was made on atom B. The reverse would follow but instead rather than taking the observation of quantity momentum one were to take instead the quantity spin i.e. the angular momentum of an atom.

This is quantised (ie occurring in specific discrete quanta). In a similar fashion if a measurement were to be made on atom A, then similarly it would already be known what that measurement would be on atom B.

Here in lies the paradox as explained in the paper. If we observe one quantity on atom A say momentum and another on atom B say spin then because of the connected nature of the two atoms it would be possible to have experimental knowledge of both so called conjugate variables.

To explain it in a table and since spin is quantised into integer quantities

atom A	atom B
expt p	$-p$
- Spin	expt Spin

atom A	atom B
expt p	$-p$
- 1	+1

[A further table directly attributable to D. BOHM would instead take the departing atoms spin components only. Without going into unnecessary details there are 3 components of the quantised angular momentum of the departing atoms. This is due to the vector nature of the quantised spin (more strictly a pseudo vector). But the same principle applies: by measuring one variable on one atom, the other is already determined on the other.

Thus one component of spin could be measured on atom A and another on B and hence full knowledge of at least 2 components of spin of A and B could be obtained.]

Thus something close to complete knowledge could be obtained of our diatomic system. This is seen as paradoxical for two reasons firstly it contradicts the Heisenberg uncertainty principle which says that there is a minimum level beyond which it is impossible to go in making definitive experimental measurements, on two so called conjugate variables.

An experimentalist can get certainty in one measurement but only at the expense of losing certainty in another. Secondly it implies that there is so called 'spooky action at a distance' according to A.Einstein ie interaction at tachyonic velocities. This is because of the virtually instantaneous connection between the two atoms deemed necessary for the two atoms to satisfy the table above as has been explained.

In our next section 2.2 the author hopes to show how Bell's theorems rises from consideration of the EPR paper and apparent paradox.

2 Discussion of Bell's Theorems

2.2(i) Introduction

As the theme of this humble work is a discussion of tachyonic physics hoping to provide a new understanding of cosmology for the 21st century the reader would perhaps not be greatly surprised if the author was to claim evidence arising from the work of John S Bell.

In Chapters 1 and 3 the author has explained that quantities having imaginary dimensions are travelling ie beyond the velocity of light. These variables are thus hidden for the simple reason that they can not be seen. What J.S. Bells contribution from the viewpoint of this author is that he showed either so called hidden variables didn't exist or if such hidden variables did exist the exhibited qualities that are described as non-local. To this author it is suggested that the **word** "non-local" means tachyonic. Effectively the word which is heavily used in the discussion of quantum mechanics

means interconnected via tachyonic velocities **in IDH space**.

2.2(ii) Discussions of hidden variables

The purpose of some hidden variables theories is to attempt to explain at an even deeper level the nature of quantum mechanics. It is an attempt to re-introduce so called determinism into physics. If there is nothing random everything can be explained from a previous development. According to Newtonian mechanics the fate of, for example a snooker ball is totally determined by the applied force. And so called hidden variables would be used to explain the apparent random movement. Occurring at the quantum level just as the movement of a snooker ball hit by something smaller say a stream of smaller objects resulting in the apparently chaotic movement of the ball as illustrated in Figure 1 below. These would be the so called hidden variables. In the actual illustrative example we could have our cue ball white and the small objects black so in a low light environment the ball would appear to be moving chaotically but if the light was turned back on, it would be apparent that the real reason for the movement of the ball was the hidden smaller objects. These would be the hidden variables and this would be described as a local hidden variable theory. It was this issue that the work of J.S. Bell was designed to address. Some of his work is contained in the book "Speakable and Unspeakable in Quantum Mechanics" referenced at the end of this chapter.

ILLUSTRATION OF LOCAL HIDDEN VARIABLE THEORY

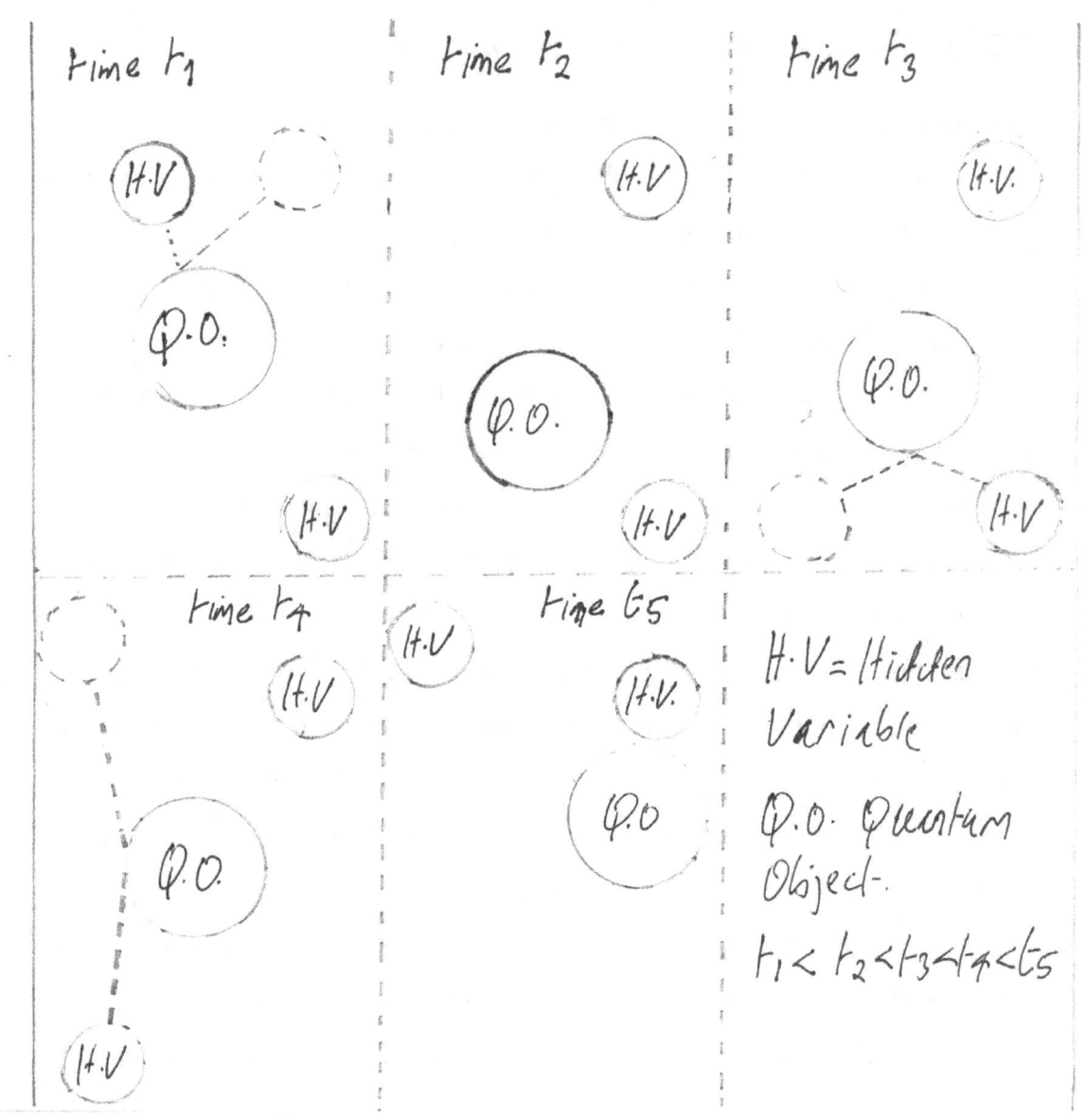

FIGURE 1. SECTION 2.

The beauty of the results which are theoretical is that experiments can be designed that can experimentally be tested. These will be discussed in the next chapter.

2.2(iii) The statement of Bell's Theorem otherwise known as Bell's Inequality.

$$\underset{(a)}{n(v,\phi_+)} + \underset{(b)}{n(\phi_+,\theta_+)} \geq \underset{(c)}{n(v,\theta_+)} \qquad \text{(A)}$$

explaining it in words, it states that if an experimentalist was to carry out 3 experiments measuring the polarizations of a large number of photon pairs in which the right and left polarizers are (a) vertical and at an angle ϕ to the horizontal (b) vertical and at an angle θ and (c) at angles ϕ on the left and θ the right, then the total number of pairs in which both photons are registered positive in the second experiment (b) must always be less than or equal to the sum of (a) and (b); subject to the assumption that the results of the experiments are determined by hidden variables possessed by the photons and *the state of either photon is unaffected by the setting of the other distant apparatus.*

It is the **italicized** statement that this author rejects.

(A paper the author is aware of by J. Clauser et al (1969) which took Bell's Inequality further by making the theoretical experiments being proposed more testable.)

An extension of Bell's Theorem states the result that

$$n(r, \phi_+) - n(v, \psi_+) + n(\theta_+, \phi_+) + n(\theta_+, \psi_+)$$

$$\leq n(\theta_+) + n(\phi_+) \qquad \text{(B)}$$

This is even more reliable as a testable equation.

These results have been stated here without much explanation but the author would expect that the reader would notice the **italicized** assumption that is contained in the explanation given here of (A).

Thus merely because most experiments demonstrate there is a violation of Bell's Inequality (A) and its extension (B) it does not in anyway rule out tachyonic travel, according to the humble understanding of this author.

REFERENCES

J.S. Bell, Speakable and Unspeakable in Quantum Mechanics, CAMB-BRIDGE UNIV. PRESS, 2nd Edition 2014.

J. Clauser et al (1969), Physical Review Letters, P.880.

A. Einstein, B. Podolsky and N. Rosen (1935) Physics Review Vol.47, P.777.

Chapter 9

DISCUSSION OF ASPECT'S EXPERIMENTS AND PROPOSALS FOR FURTHER RELATED EXPERIMENTS

1 Introduction

Some of the key experiments relating to our new cosmology the author is setting out in this hopefully relatively short book were carried out by ALAIN ASPECT (Ref at the end of the CHAPTER).

Because the experiments were laboratory based experiments and for other reasons it was felt necessary to confirm the results over a greater distance. These were carried out under guidance of A. Zeilinger on a major river passing through NOVYSAD (Translated NEW GARDEN) the DANUBE, at VIENNA.

Thus the A.ZEILINGER experiments gave greater experimental confirmation to Aspect's key results. As far as the author is aware his proposal that would provide a tentative experimental check in the principles of tachyonic physics has not yet been carried out, but if such experiments have been carried out they would be likely confirm this work.

2 DISCUSSION OF ASPECTS EXPERIMENTS

Aspect's experiments provide an experimental test on the theoretical predictions provided by the various versions of Bell's inequality mentioned in the previous chapter.

They confirm the correlation expected between the two different polarisation locations and thus violate Bell's inequality.

According to the authors understanding this is entirely to be expected. This is because the two polarisation locations are fixed in the laboratory as illustrated below and thus their velocity is zero relative to each other.

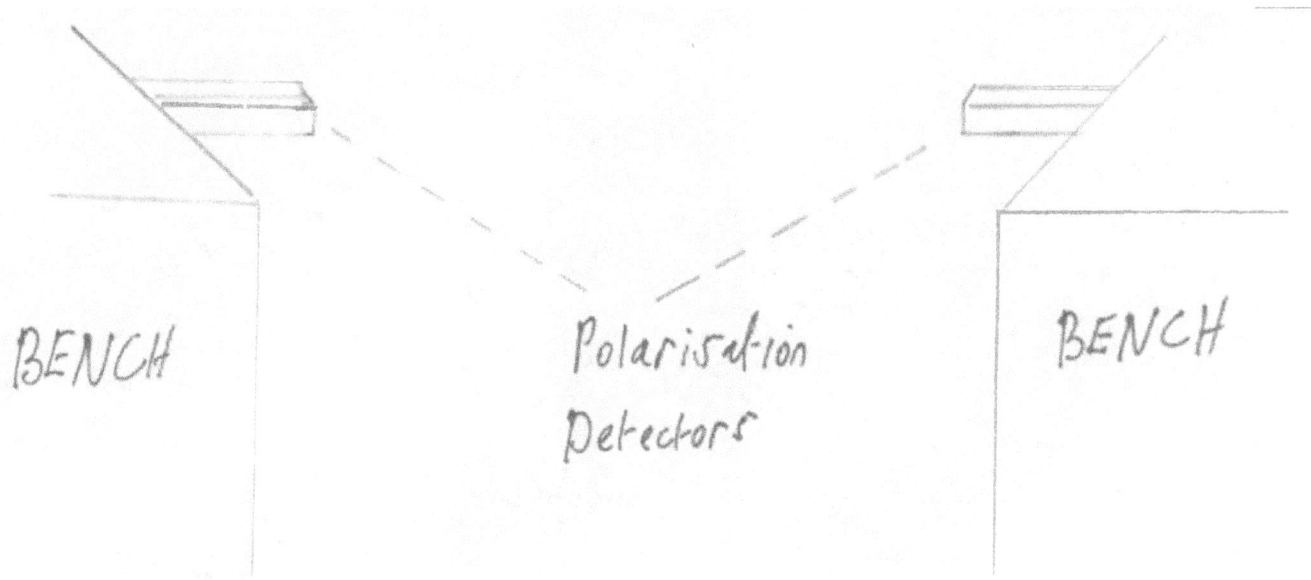

DIAGRAM ILLUSTRATING ASPECTS FIGURE 1. POLARISATION DETECTION DIAGRAM.

ILLUSTRATION OF AN ALTERNATIVE EXPERIMENT.

MEASUREMENT APPARATUS FOR CONJUGATE VARIABLE TWO

DIATOMIC MOLECULE

MEASUREMENT APPARATUS FOR CONJUGATE VARIABLE ONE.

MEASUREMENT APPARATUS INITIALLY STATIONARY

MOLECULE SPLITS INTO

TWO ATOMS MOVING APART

MEASUREMENT APPARATUS INITIALLY STATIONARY.

MEASUREMENT APPARATUS MOVES

A

B

TOWARDS OTHER MEASUREMENT APPARATUS

MEASUREMENT IS MADE ON ATOM A OF VARIABLE TWO

MEASUREMENT IS MADE ON ATOM B OF VARIABLE ONE

58

To explain again according to our $\frac{C^2}{V}$ result again if $v \to 0$ then $\frac{C^2}{V} \to \infty$ and thus the correlation would be instantaneous.

However if the polarisation detectors were to move with velocity V relative to each other then we would expect a different result. The $\frac{C^2}{V}$ would apply. (This explains why there are limits to the degree of so called quantum entanglement.) Quantum entanglement is the term that is used by some author to explain the correlation that is observed between distant parts of a previously connected whole such as a diatomic molecule, the example mentioned in CHAPTER 8.

3 DISCUSSION OF THE DIATOMIC MOLECULE EXAMPLE GIVEN IN CHAPTER 8

As was mentioned in CHAPTER 8 the example that was used to illustrate the so called EPR paradox was provided by DAVID BOHM.

Where as the Aspect Experiments and their extension to the Zeilinger experiments related to light and its polarisation states there also exists the potential for an experiment with ordinary matter, so called tardonic matter ($V < C$) using the diatomic molecule example given in CHAPTER 8.

Suppose our DIATOMIC molecule were to split and our experimentalist were to measure the two conjugate variables as explained in CHAPTER 8.

At first sight the obvious experiment would be to have our two separate measurement apparatuses fixed within the laboratory however far or close they would be to each other.

Such an experiment would not be any different in principle to Aspects experiments or their Zeilinger extensions in the sense that the two experimental apparatuses would be instantaneously connected tachyonically due to their zero relative velocity.

However suppose the two apparatuses were to move relative to each other then this author suggests different results would be obtained.

In this situation the tachyonic connection between the two experimental setups occurs at velocity $\frac{C^2}{V}$.

REFERENCES

ALAIN ASPECTS p119-53 in Quantum Unspeakables: From Bell to Quantum Information

BERLIN: Springer-Verlag 2002. Edited by R. Bertlmann and A. Zeilinger.

ILLUSTRATION OF AN ALTERNATIVE EXPERIMENT.

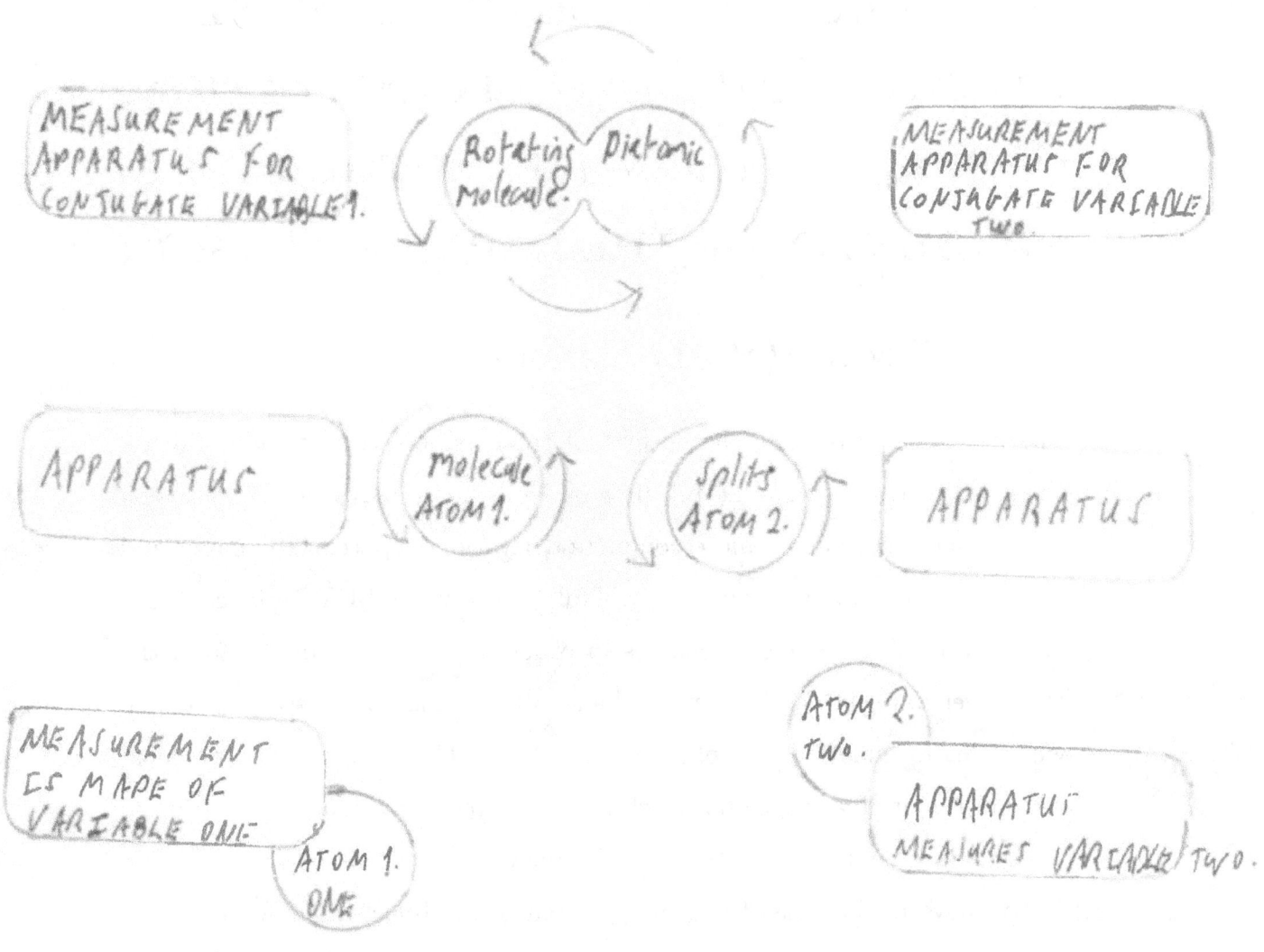

FIGURE 3. DIAGRAM.

Chapter 10

DISCUSSION ON TACHYONIC MECHAN-ICS TWO ALTERNATIVE FORMULATIONS ONE OBTAINED BY THE AUTHOR INDE-PENDENTLY OF THE EARLIER VERSION

1 The R.A.S. (1991) Version

This version is published in the first of three papers two of which are reproduced in this book. The full first paper is not reproduced in this book because chapter one is a more complete derivation of part of the paper and for reasons of brevity it would be useful to study the SUTCLIFFE (1991) formulation of tachyonic mechanics in order to see how it differs from the earlier version (1967). The author had no knowledge of the earlier version when this formulation was obtained in 1991. (There is one typographical error in the original page 160 which is now corrected).

Quoting directly from pages 160 the relevant equations.

We have in the case of the energy momentum 4-vector the following relations

$$E = \frac{m_0 C^2}{\sqrt{1 - \frac{V^2}{C^2}}} \qquad (1)$$

$$P = \frac{m_0 V}{\sqrt{1 - \frac{V^2}{C^2}}} \qquad (2)$$

If we imaginify the left hand side, if we are to be consistent, E will become iE and P, iP.

However, this will mean that on imaginification the scalar rest must change its sign because the imaginary quantity on the right hand side is on the bottom line (denomination) and $\frac{1}{i} = -i$.

Thus (1) and (2) becomes:

$$iE = \frac{-m_0 C^2}{\sqrt{1 - \frac{C^2}{V^2}}} \qquad (3)$$

$$iP = \frac{-m_0 \frac{C^2}{V}}{\sqrt{1 - \frac{C^2}{V^2}}} \qquad (4)$$

On substituting from (3) and (4) into the well known relationship $\frac{E^2}{C^2} - (P)^2 = m_0^2 C^2$. N.B. In (4) when $V = 0$ the result is indeterminate. We obtain

$$\left(\frac{iE}{C}\right)^2 - (ip)^2 = -(-m_0)^2 C^2.$$

The tackyonic energy, momentum relationship, which can be rewritten $(iPC)^2 - (iE)^2 = M_0^2 C^4$, the R.H.S. being the invariant. (3) $\div$ (4) $\frac{iE}{iP} = V.$ $E. = P.V$. This relationship holds for photons and tackyons.

2 The U.S. Version

There are subtle differences between the two versions. The author came across this version in WIKAPEDIA early in 2014 and considers it worthy of inclusion so a reader can have a sense of how slow advances are made in the understanding of this developing area of physics.

Gerald Feinberg (1967) proposed a different more restricted version of tachyonic mechanics but it is included here for the sake of completeness. G. Feinberg (deceased 1992). G. Feinberg published his ideas formally aged 34.

Feinberg proposes that *the total energy must be real in tachyonic particles F(1)*.

$$E^2 = p^2 C^2 + m_0^2 C^4 \qquad F(1)$$

$$E = \frac{m_0 C^2}{\sqrt{1 - V^2/C^2}} \qquad (2)$$

Hence using the above equation (2) Feinberg suggests that $V > C, m_0$ the mass of the particle is imaginery subject to $F(1)$ a relatively straightforward statement to see. (An earlier paper by BILANIUK et (1962) used the term meta particle. Meta relatively American Journal of Physics 30 (**10**) page 718.)

(Also Feinberg 1978 Physics Review **P** 17 198.)

3 A DISCUSSION OF THE ELECTROME-CHANICAL TACHYONIC RELATIONSHIPS

"By using the same mathematical technique on the current, charge density 4-vector namely the imaginification process the author obtained the following results.

Similar considerations apply to the current, charge density 4-vector.

We have that

$$J = \frac{\rho_0 V}{\sqrt{1-\frac{V^2}{C^2}}} \quad (5) \quad \text{and} \quad \rho = \frac{\rho_0}{\sqrt{1-\frac{V^2}{C^2}}} \quad (6)$$

(5) current density   (6) charge density

On imaginification we have that

$$iJ = \frac{-\rho_0 \frac{C^2}{V}}{\sqrt{1 - \frac{C^2}{V^2}}} \quad i\rho = \frac{-\rho_0}{\sqrt{1 - \frac{C^2}{V^2}}}$$

the sign being thought of as a change in the sign of the charge on imaginification which is necessary to maintain consistency.

The reverse process to imaginification would thus involve a change in the sign of the charge and mass from negative to positive and we thus justify the use of the term positivisation".

The mass and charge densities attached to tachyonic particles can thus be thought of as scalar labels associated with the particles which change their sign on imaginification and change it back again on positivisation.

Directly quoted from SUTCLIFFE (1991), Published (1994) Page 161. One typographical error was put in the original version namely "if" should be "of".

The above quoted equations are an extension of the imaginification concept created by the author and first published on the scientific record in the conference proceeding of the 1991 R.A.S. conference on problems in space, time and gravitation. To the knowledge of the author Feinberg did not apply his ideas to the electromechanical relationships merely using equation (1) Section 2 of this chapter as his starting point.

By contrast the author extends his formulation to the electromechanical relationships thereby reaching his conclusion as to the **change** of sign of the **charge** in the process, similarly to the mass. Another obvious difference between the two versions is that while the author considers the mass of the tachyonic particle to be real (and negative) while the energy is imaginary Feinberg makes the claim in $F(1)$ that the energy must be real and thus the mass is imaginary.

They are alternative ways of looking at the same idea. However the author asserts the mass is the more fundamental identifier in conventional S.I. physics and as such it makes sense for the mass to be real and the energy to be imaginary. The author also asserts that his version is more comprehensive because it applies to charge as well as other fundamental quantities in the S.I. or c.g.s. measurement system.

However the author freely acknowledges that G. Feinberg (deceased) was thinking along the same lines as the author at a far earlier period. Perhaps G. Feinberg wasn't taken as seriously as the author because he didn't explain his ideas to a Russian or more international audience, and perhaps he was interested in other areas of physics.

Because the author starts from his imaginification concept and its reverse positivisation concept he is able to use this conceptual framework **along** with another proposal in CHAPTER 4 to outline a development **of tachyonic** quantum mechanics, that is set out in CHAPTER 3 **and Appendix.** The author effectively re-interprets and redevelops quantum mechanics using tachyonic concepts, where as FEINBERG doesn't expand his interpretation in his way.

REFERENCES

GERALD FEINBERG (1967) PHYSICAL REVIEW (159)5. Page 1089-1105.

M.H. Sutcliffe p152-162 Some Observations Concerning Velocities Greater than the Velocity of Light. Conference proceedings R.A.S. 2nd International Conference on Problems in Space, Time and Gravitation held in St Petersbourg Sept 16-21st 1991.

Chapter 11

A SPECULATION AS TO THE NATURE OF DARK ENERGY AND DARK MATTER, TACHYONIC GRAVITY

11.1 INTRODUCTION

A reader having followed the ideas within this book would not be surprised to learn that this author would suggest that the nature of the unexplained energy within the cosmos would have something to do with tachyonic physics.

If this author is correct in his understanding of DARK ENERGY and DARK MATTER it would appear there is some missing energy within the universe under current scientific thinking. Using the logic of PARMENIDES there are two possibilities.

Possibility 1. The current idea of there being some unexplained ie missing DARK ENERGY is wrong or

Possibility 2. It is correct.

If Possibility 2 is correct then we have to explain where the DARK ENERGY/DARK MATTER is?

Surely one possible speculation is that this DARK ENERGY/DARK MATTER is tachyonic in nature.

This explains why it cannot be seen and observed by scientists using current scientific methods.

11.2 FOUR NEW VERSIONS OF GRAVITY

DARK ENERGY OR DARK MATTER WOULD HAVE SOME GRAVITA-
TIONAL EFFECT, IF WE MAKE THE SOMEWHAT BOLD ASSUMP-
TION THAT THE LAWS of PHYSICS APPLY TO BOTH SIDES OF THE
LIGHT BARRIER.

Since the basis of this book which is only the beginning of a serious effort
to reformulate physics and provide a new cosmological understanding for our
time is based on *that assumption*, but also being fully aware that much further work will be required as is the nature of modern physics, let us make
an effort to at least partially explain the gravitational effect of DARK EN-
ERGY/DARK MATTER USING NEWTONS LAW OF GRAVITATIONAL
ATTRACTION.

Our reader may be familiar with this fundamental law of physics which
GENERAL RELATIVELY THEORY explains using the principle of Equivalence. However because this author has questioned some of Einsteins work
he feels it easier to work with the more established Newtonian law and see
where it takes us.

Newtons LAW OF GRAVITATIONAL ATTRACTION

$$(1) \qquad F = \frac{GM_1M_2}{r^2}$$

G = Newtons Gravitational constant

$M_1 =$ MASS ONE

$M_2 =$ MASS TWO

$r =$ distance between the centre of gravity of each mass.

TWO EQUATIONS ARISING FROM CHAPTER 10 OF TACKYONIC GRAVITY

In the SUTCLIFFE version set out in CHAPTER TEN when we imaginify, our masses change sign

$M_1 \to -M_1$

$M_2 \to -M_2$

$r \to ir.$

This is the imaginification process.

(The reverse process of positivisation gets us back to the matter.)

THIS YIELDS ONE EQUATION OF TACHYONIC GRAVITY. BASED ON AN ASSUMPTION THAT WILL BE EXPOSED.

Thus

$$F = \frac{G(-M_1)(-M_2)}{(ir)(ir)}$$

$$= -\frac{GM_1 M_2}{r^2} \text{....Sut (2)}$$

This is a mathematically real result.

However it is based on the assumption that the Gravitational constant G doesn't change, on imaginification.

This would lead to apparent contradictions but is included because it may still have some validity.

Thus to maintain consistancy in our approach on imaginification $G \rightarrow iG$.

Hence Sutcliffe (3) the imaginification of the Newtonian LAW OF GRAVITATIONAL ATTRACTION obtained from (2) by replacing G by iG is

$$F = -\frac{iGM_1M_2}{r^2} \qquad \text{...Sut (3)}$$

Thus (3) is an imaginary force which is opposite in sign to (1) just as equation (2) is.

This explains DARK MATTER in a simple fashion because if we use the idea of what the I.P.P. is from CHAPTER 3 and 4 then the dark matter is **not attracting** in tachyonic space.

FEINBERGS TWO VERSIONS.

Our reader who has followed this text to its soon to be concluded end will note that FEINBERG CHAPTER TEN also used a different but similar approach, so this is included for completeness:

starting with equation (1)

$m \rightarrow im$

substituting into (1)

$r \rightarrow ir$

$$F = \frac{G(im)(im)}{(ir)(ir)} \qquad \text{which merely...} \qquad \text{(Feinberg (2)}$$

gets us back to (1) (the ({i})s cancel top and bottom.) NEWTONS LAW.

However if we allow Feinberg's version to use a real r we end up with (3).

$$\text{ie.} \quad \frac{G(im) \times (im)}{r \times r} = -\frac{-Gm^2}{r^2} ie \quad \text{(Feinberg (3)}$$

but this involves mathematical trickery because distance is clearly imaginary/complex in the tachyonic part of the cosmos as is explained in CHAPTER ONE AND THREE. (This does not alter the fact that Feinbergs version and this authors version are remarkably similar even though they started in different places, the author starting from the fundamental space-time relationships and Feinberg starting from the energy equation, mentioned in the previous chapter).

The effect of either Sut(2)or(3) would be to create a gravitational repulsion. The author doubts (2) but includes its derivation for completeness of presentation, but has far greater confidence in (3) which would also show that masses in the tachyonic aspect of the cosmos are moving away from each other. [On positivisation they would then be further apart and this rate would be accelerating.]

The fact that Feinberg (2) merely returns us to our starting equation of Chapter 10, equation (1) should serve to convince a reader that this later version and presentation of tachyonic physics seems more creditable, and is more comprehensive than that of Feinberg.

This concludes this speculative chapter which explains at last partially the mysterious missing dark matter.

Chapter 12

CONCLUSIONS

The purpose of this book is to look in an alternative way at some of the well established ideas from quantum mechanics and to try to link them to tachyonic concepts thereby offering the possibility of a new cosmological outlook for the 21st century and beyond. This is an outlook that is characterised by a refutation of the frequently made claim in various 20th century physics texts that 'nothing can go faster than the speed of light'. Once this dogma has been set aside areas of research previously off limits open up. And this for the most part involves re-examining how the equations of quantum mechanics are being derived and interpreted.

In chapter three the author outlines a possible derivation of DIRACs formulation of quantum mechanics from tachyonic concepts. In chapter ten the author points to the similarity between his version of tachyonic mechanics and that of Feinberg.

They were derived in completely different ways and yet are remarkably similar. This author was completely unaware of the name Feinberg or his work up until 2014 and yet feels it is essential Feinberg is mentioned.

The author invents the concept of imaginification and its reverse process positivisation from within the Lorentz transformation and derives V_C which is then described as the corresponding velocity. But then having derived the result (ie of V_C) the author notes that this result is already contained within the foundations of quantum mechanics. There are two different derivations one within the key paper by Louis de Broglie and the other by Paul DIRAC. When the author first developed this result he found it curious if not astounding.

How would it be that using S.R.T. one could end up with a result that is found within the framework of quantum mechanics. The derivation is obtainable by straightforward methods and is reproduced in full in CHAPTER ONE. This is to enable the reader to check there are no errors within it, and to show there are no further assumptions than the ones set out. The author examines the original Einsteinian proof that is translated into English and is found in W. PAULIS book.

This book, and its proof is clearly flawed within the understanding of this author but once again what is curious is it leads to the universal corresponding velocity V_C, indirectly, as is shown in CHAPTER ONE with an inequality argument.

It is this which has lead the author to suggest that since the proof is flawed then perhaps the alternative being offered in this new cosmological outlook is worth considering.

The reason why this cosmological outlook is appropriate to the 21st century relates to the development of science and technology since the time of I.Newton although this author also references an ancient, Parmenides who invented some of the key ideas of western science and logic, but is less understood and known than **Isaac** Newton.

Newtonian science emphasises mechanical force. The Manchester scientist James Joule emphasised energy and showed it was conserved, in the 19th century. Developments in steam engines and motors were accompanied by increasing understanding about electricity both theoretically and experimentally. The experimental work of Michael FARADAY from **Lancashire**, ENGLAND and the Scottish born mathematical physicist J.C. MAXWELL, spring to mind, but many others contributed. The concept of wattage, the electrical power, the rate of using electrical energy, is named after JAMES WATT, because the electrical use of energy as measured is considered the same as the mechanical use of energy in for example the steam engine, invented by WATT.

All these understandings about electrical power and its equating with mechanical power, energy and force were developed from the time of Newton using classical mathematics and physics into the 19th century. And they related to the industrialisation occurring at those times.

These processes of technological advance and theoretical advance primarily in the use and generation of energy continued in the 20th century. But there is one more concept that a reader will perhaps be aware of that this author sees as important because of the way SHANNON and KHINCHIN very intelligently reinterpreted it, that arose in the 19th century. That is the concept of Entropy. Entropy is a very curious concept indeed, and one that has taken on an important role in the 21st century. Let me explain. Entropy is a measure of the disorder of a closed system. Two systems can have the same energy but different entropies. Entropy increases as time moves forward within a closed system. (ILYA PRIGOGINE emphasis that this is not necessarily the case within an open ie interacting system.) But the key idea arising from the entropy concept is the loss of Entropy is the gain of information ie our reader will perhaps understand how in some intuitive sense this makes sense, even if their grasp of this particular aspect of mathematics is not too strong.

The typical example quoted in the literature is the cracking of the egg, for example onto a frying pan. The cracked egg is more disordered and hence its entropy has increased. (Unless heated up both states of the egg contain the same digestable energy ie Joules.) Another way of explaining it is to say that the uncracked egg has higher information content, the cracked egg less.

This is a nutshell (no pun intended) is the intelligent way that SHANNON and KHINCHIN defined and developed the mathematical theory of information.

In his original (1991) paper the author made the significant observation that there is nothing in the mathematical definition of observation that limits the transmission of information to velocities less than C.

The invention of the first electronic computer by Professor FREDERICK C WILLIAMS and his student TOM KILBURN later Professor KILBURN, of MANCHESTER UNIVERSITY, that operated a computer program marked, the beginning of the information age.

In the 21st century, the emphasis will be less on the raw use of energy and more on the intelligent use of information, to make advances in science, to use energy more wisely.

This is particularly the case when we take into account the findings of climate science, this is discussed for example in a work by Dr ROSALIE BERTELL. This is why the cosmology outlined here is appropriate to our time.

There are some ideas the reader may find interesting in CHAPTER 3 in addition to the tachyonic derivation of quantum mechanics. Namely the holographic nature of the universe showing its interconnected nature and the 4-force 4 dimension isomorphism theorem. All this theorem does is it equates time with gravity and the other three forces with this 3 dimensions of space. The three forces ie short nuclear, long nuclear and electromagnetics force are already unified under an existing physics theory. Perhaps this idea may help to further advance physics.

CHAPTER 11 contains a derivation of tachyonic gravity and is an original derivation **obtained** by the author **in 2015** which the author speculates **explains Dark Energy,Dark Matter and the expanding cosmos.** This concludes this book.

REFERENCES

R.BERTELL PLANET EARTH 2000 THE WOMEN PRESS Ltd. PAGES 136-139 ISBN 0704346508.

W. PAULI THEORY OF RELATIVITY, DOVER PUBLICATIONS 1958 Edition. ISBN10 048664152X.

I. PRIGOGINE, STENGERS ISABELLE (1984). Order out of CHAOS Man's New dialogue with nature., Flamingo ISBN 0-00-654115-1.

GLOSSARY OF TERMS IN TACHYONIC PHYSICS

FEINBERG VERSION:

VERSION OF TACHYONIC MECHANICS PUBLISHED BY G FEINBERG.

I.D.H. pace: Infinite dimensional hyperspace; the space of tachyonic physics.

IMAGINIFICATION: Concept proposed by the author to explain the transformation from one side of the light barrier to the other. This provides the initial conceptual framework for tachyonic physics, where as Feinberg started from the energy relationship the author started from the more fundamental space-time conceptual relationship.

POSITIVISATION: The reverse process to imaginification. Positivisation means passing from the tachyonic (faster than light) side of the light barrier to the slower (tardonic) side of the light barrier.

TACHOR/TACKOR: A vector representing a tachyonic state.

TACKYONIC: Alternative spelling of TACHYONIC. TACHYONIC: Faster than the velocity of light.

TACKYONIC GRAVITY: A derivation of Newtons Law applied to tachyonic physics by the author. Provides a possible explanation of dark energy.

SPACE-TIME ISOMORPHISM CONJECTURE. A simple idea equating the four known forces with space-time, GRAVITY being equated with time.

Appendix A1

(THIS PAPER EXTRACTED FROM A 1988 UNPUBLISHED PAPER.)

Additional Explanation of the TACHYONIC DERIVATION OF QUANTUM MECHANICS

THE SIGNIFICANCE OF LINEAR OPERATORS.

Then we make the assumption that the imaginification and positivisation operations occur conversely. They are effectively linear operators. These have the effect of leaving the system unchanged.

When we make a measurement at the specific co-ordinate point (x, t) in one tardonic frame we extract a real number from the tachyonic state of the eigenvector which describes the corresponding point in I.D.H. space. This involves the Positivisation operator. Simultaneously at the point (x, t) the Imaginification operator is extracting the eigenvector from the vector (x, t) in 4 space.

Thus the operation of the linear operator (see remark 5H (as below)) produces real information which is all obtained from I.D.H. space (which is of an imaginary nature)(strictly the field is complex but the vectors all have an imaginary factor in all co-ordinates because of the method of their formulation).

$Itc \rightarrow td$ is the symbol for the positivisation operator in the proof below

REMARK 5H Proof of linearity for a specified time t^1 of the positivisation operator

$$i(x_1^1 + \alpha x_2^1) \quad Itc \rightarrow td = ix_1^1 Itc \rightarrow td + \alpha(ix_2^1)$$

$$Itc \rightarrow td(\forall t^1, V_c)\alpha \in C(C \text{ complex number field}).$$

similarly fixing x^1 we can do the same positivisation for time and it is similarly linear.

Consider the mathematical theorem of Analysis that from the point of view of the real no.s it is impossible to order the complex no.s in order of increasing size. Thus from my point of view in real 4-space the particular value on the other side of the light barrier is insignificant. It is only when positivisation occurs that the magnitude of the tardonic physical quantity becomes important.

It is reasonable to assume that throughout 4-space a particular constant of proportionality will occur for all the different I.D.H. co-ordinates that exist on positivisation. We would expect this value at each point to be small because it is only referring to one point.

Suppose we make a measurement of the eigenvector at the specific time

t_1 i.e. we positivise with respect to frames K^1 and K.

$$| p >= \begin{bmatrix} 0 \\ 0 \\ 0 \\ . \\ . \\ . \\ i_1 x_1 \\ 0 \\ 0 \\ 0 \\ . \\ . \\ . \\ 0 \end{bmatrix}$$

Suppose $v = 0$, $VC = VI = \infty$ $K = K^1$

A linear operator can be formed by taking the operator $\frac{\partial}{\partial i x_1}$.

Let us call the constant of proportionality h. This constant is according to our assumptions valid throughout 4 space.

The operation of the operator on the vector will produce an eigenvalue say $\mathbf{P}x_1$.

Thus we get

$$h\frac{\partial}{\partial ix_1}\begin{bmatrix} 0 \\ 0 \\ 0 \\ \cdot \\ \cdot \\ \cdot \\ ix_1 \\ 0 \\ 0 \\ 0 \\ 0 \end{bmatrix} = \mathbf{P}x_1\begin{bmatrix} 0 \\ 0 \\ 0 \\ \cdot \\ \cdot \\ \cdot \\ ix_1 \\ 0 \\ 0 \\ 0 \\ 0 \end{bmatrix}$$

N.B. remember that the zeros in the eigenvector are arbitrary but irrelevant to a partial derivative of a particular co-ordinate.

But

$$\frac{\partial}{\partial ix_1} = \frac{\partial}{i\partial x_1}.$$

This implies that

$$\frac{h}{i}\frac{\partial}{\partial x_1}|p> = \mathbf{P}x_1|p> \quad ...(A)$$

The eigenvalue is real and thus is a measurement in 4 space. For intuitive reasons we assume that h is Planks constant (as found in the well known $E = h\mu$ where $c = \lambda\mu$) divided by "π.

For reasons of dimensionality $\mathbf{P}_x$, thus is a momentum measurement on the quantum level.

The other axis of our 4 vector is time and we have a similar equation

$$-h\frac{\partial}{\partial it_1}\begin{bmatrix} 0 \\ 0 \\ 0 \\ 0 \\ 0 \\ it_1 \\ 0 \\ 0 \\ 0 \\ 0 \end{bmatrix} = Et_1\begin{bmatrix} 0 \\ 0 \\ 0 \\ 0 \\ 0 \\ it_1 \\ 0 \\ 0 \\ 0 \\ 0 \end{bmatrix}$$

or

$$-\frac{h}{i}\frac{\partial}{\partial t_1}|E> = Et_1|E> \quad \ldots\ldots\ldots(B)$$

We assume that our tachyonic vector or tachor is being positivised at a fixed x_1 position. We use the same constant of proportionality h because we are putting all dimensions in 4 space on the same footing.

We choose the $-ve$ sign carefully. Clearly due to dimensionality Et_1 is an energy and it must be positive.

We recall in our considerations we have referred to two frames VK and VK^1.

It is assumed that positivisation occurs in frames VK^1 and this is observed from VK. However the above equations (A) and (B) refer only to one frame response. So we are interested in the case where VK becomes identical with VK^1. In this case the rate of change of distance $\dfrac{\partial}{\partial x_1}$ as $v \to 0$

83

is positive. However the rate of change of time $\dfrac{\partial}{\partial t_1}$ as $v \to 0$ is negative. Thus we choose to place a negative sign in front of it.

We realise that there may be and are other linear operators that acting on I.D.H. space produce real no.s and some of those will correspond to other measured quantities. Essentially the act of positivisation corresponds to the act of measurement which means the extraction of a real quantity from an eigenvector.

THE FORMULATION OF THE STATE VECTOR.

We have to be extremely careful in our thinking to understand the problem of the quantum measurement of a system and the related problem of the formulation of the state of the system.

As we have made clear in our study of eigenvectors a displacement eigenvector is defined at a particular time.

A measurement system at a particular time essentially is a co-ordinate frame in 4 space. All the possible co-ordinates at that time determine the nature of our tardonic frame of reference. For simplicity we assume as before we are using the Schrodinger representations which correspond to the space-time 4 vector in 4 space and that the y, z axis co-ordinates are set equal to zero in the manner of a space-time diagram.

When we make a measurement of momentum at a particular place; x_1^1 on the x^1-axis we have specified an eigenvector already. However if we measure momentum at a general place on the x^1-axis we allow the measurement process to determine where we measure the momentum of a particular particle say an electron. As the point of measurement is undefined we have to consider all x_i^1 positions. Thus the vector representing the state of the measurement system is the sum of all the infinite number of possibilities of eivenvectors. Thus the state of the system is the following sum which we call the state vector $|\psi t_0 >$

$$|\psi t_0 >= \begin{bmatrix} -i\infty \\ 0 \\ 0 \\ 0 \\ 0 \\ 0 \\ 0 \\ \cdot \\ \cdot \\ \cdot \\ \cdot \end{bmatrix} + \ldots \begin{bmatrix} \cdot \\ \cdot \\ \cdot \\ ix_1^1 \\ 0 \\ 0 \\ 0 \\ \cdot \\ \cdot \\ \cdot \\ \cdot \end{bmatrix} + \begin{bmatrix} \cdot \\ \cdot \\ \cdot \\ 0 \\ ix_1^1 \\ 0 \\ 0 \\ \cdot \\ \cdot \\ \cdot \\ \cdot \end{bmatrix} + \ldots \begin{bmatrix} 0 \\ 0 \\ 0 \\ \cdot \\ \cdot \\ \cdot \\ \cdot \\ \cdot \\ \cdot \\ \cdot \\ i\infty \end{bmatrix}$$

N.B. if $K = K^1$

$$x_1 = x_2 = x_3 = \cdots \infty.$$

This sum is an infinite sum and each vector has an uncountably infinite number of entries.

Such a sum would normally be expressed as an integral $|\psi t_0 >= \int |\epsilon > d\epsilon$ and $|\epsilon >$ is an eigenvector of displacement. When a measurement is made effectively one of the eigenvectors is selected. However this basically means the specifying of the x_i^1 general coordinate to a particular value, say x_1^1. This will be done relative to the measurement apparatus which determines the co-ordinate frame. This corresponds to our co-ordinate frame K^1. Thus relative to another aspect of the measurement apparatus the frame K, the $|\psi t_0 >$ vector moves with velocity Vc. And relative to the frame K^1 the $|\psi t_0 >$ vector moves with velocity VI. If we might consider two tardonic frames as indistinguishable then $v = 0$ and the velocity of the vector $|\psi t_0 >$ is infinite. This means that $|\psi t_0 >$ pervades all the I.D.H. space that corresponds to our chosen 4-space.

N.B.Note that this state vector $\psi t_0 >$ as defined above is purely a state vector obtained by considering a measurement at a particular time. We thus call it particular time state vector. It is not an eigenvector in general.

THE GENERAL STATE VECTOR

This will be the sum of all previously existing particular time state vectors for our chosen frame of reference K^1

i.e. $|\psi >= |\psi t_1 > +|\psi t_2 > + \cdot +|\psi t_0 >$ here we call our particular time t_0, where $t_1, t_2, \cdot, t_0$ are times when the chosen frame of reference K^1 and its measurement has existed.

However a measurement apparatus has only existed when measurements have been made. Thus $|\psi t_i > i \neq 0$ are effectively eigenvectors. Measurements are made when the concurrent process of imaginification and positivisation have occurred. It is only the particular time state vector better named the present time state vector that is not an eigenvector. All the other such vectors are eigenvectors that have been collapsed to specify a measurement or eigenvalue.

When a measurement is made at a general time, the positivisation operator effectively collapses the $|\psi >$ vector to get a particular eigenvalue. This will depend on the what is being measured. Clearly from the definition of the $|\psi >$ vector above the value can be either an eigenvalue that has already been in existence at time $t_i i \neq 0$. Or it can be a new eigenvalue that hasn't already been in existence but can be extracted from the present time or particular time state vector.

This will lead to a range if values for our measurement with far greater probability values occurring for the values that have previously occurred most frequently. This provides the basis for the probabilistic interpretation of Quantum mechanics.

We make the assumption that each $|\psi t_i >$ has the same status as every other. The prob of obtaining value α say is given by

$$\text{prob}(\alpha) = \frac{\text{no. of eigenvectors that have produced value } \alpha}{(\text{total no. of eigen vectors} + 1)} +$$

$$\frac{[\text{total number of ways of obtaining } \alpha \text{ from } |\psi t_0 >] \times \beta}{[\text{total no. eigenvalues extractable from } |\psi t_0 >]}$$

$$\beta = \frac{1}{(\text{total no. of eigenvectors} + 1}$$

MOMENTUM REPRESENTATION.

So far we have worked with the 4 vector (x, y, z, t) although we have largely ignored y, z values for simplicity.

If instead we work with the 4 vector $(P_x, P_y, P_z, \frac{E}{c})$ all the algebra looks just the same. However our eigenvectors will be momentum eigenvectors. Our particular time will become a particular energy. We make the same assumption relating the projection from I.D.H. space to 4 space concerning the constant h that doesn't concern itself with particular imaginary co-ordinates

$$-\frac{\partial}{\partial iP_i}h = ih\frac{\partial}{\partial p_i} \quad \text{has as eigenvalues the}$$

value q_i for our eigenvector $|P>$.

By Dimensional Analysis the value q_i is a distance measurement. It represents the distance at a particular energy value that you will find the positivised particular value with the chosen eigenvector.

We can also form as eigenvector at a particular momentum value. We choose to define our linear operator in terms of energy rather than the energy divided by c.

This is because c is a constant (as a strong approximation with the local frame) and thus if we specify Energy we specify E/c for any value of E.

$$-\frac{\partial}{\partial iEx}h \begin{bmatrix} 0 \\ 0 \\ 0 \\ iEx \\ 0 \\ 0 \\ 0 \\ 0 \end{bmatrix} = t \begin{bmatrix} 0 \\ 0 \\ 0 \\ iEx \\ 0 \\ 0 \\ 0 \\ 0 \end{bmatrix} \quad \text{for a particular value } px \text{ of } p.$$

By Dimensional analysis the eigenvalue is time.

$$-\frac{h}{i}\frac{\partial}{\partial E_x}|E> = t|E>.$$

General 4-vector analysis.

We have now used two four vectors the displacement time (x, x_2, x_3, t) and the $(P_1, P_2, P_3, \frac{e}{c})$ 4 vector. By implication we can proceed in a similar manner to use other 4 vectors. The analysis will proceed in the same manner and will we assume involve the same constant of proportionality when we formulate our linear operator from the partial derivatives. We will expect the quantity corresponding to time, T, to have the operator $-\frac{\partial}{\partial iTx}h$ associated with it and we will expect the value of the quantity corresponding to distance, l, to have the operator $\frac{\partial}{\partial iLx}h$ associated with it. When we use the words "corresponding to" we mean "satisfying the same aspect of the Lorentz transformation as".

THE SCHRODINGER EQUATION - SOME OBSERVATIONS

We have the eigenvector equation $-\frac{h}{i}\frac{\partial}{\partial t_i}|E> = Et_i|E> (\forall i)$

This equation is for a particular fixed x, position on the x-axis at a particular time t_i. The positivisation is occurring in a particular frame of reference K^1 which is tardonic. WHile the $|E>$ vector exists in a tackyonic frame of reference.

It is important to remember that if $V = 0$ then our observer frame $K = K^1$. Additionally the vectors $|E>$ in I.D.H. space travel at infinite speed and effectively pervade I.D.H. space. This means that at $\forall$ time, we can collapse our general state vector $|\psi>$ to get a particular eigenvector. This will be done in the frame K^1 and observed from our form K. These will be the same if $v = 0$.

As the eigenvectors pervade I.D.H. space we could collapse the eigenvectors at a number of places in K^1. The resulting eigenvalue represents the energy at the separate places at which the general state vector is positivised. We call the total energy the Hamiltonian.

Thus the general state vector will progress in time in accordance with the law $-\dfrac{h}{i}\dfrac{\partial}{\partial ti}|\psi >= Ht_i|\psi >$ when Ht_i is the total energy considered in K^1 at time t_i.

This is the so called Schrodinger Equation. It gives the progression of the state vector in the frame K^1. And as repeatedly stated if $v = 0$ then $K = K^1$ (Newtonian estimates are valid for small v.)

Perhaps the author will derive the Dirac equation in a later edition of this book. This involves making assumptions about the internal degrees of freedom of particles. The so called spin properties of an electron.

The machinary of Quantum mechanics that we have developed ignores the internal degrees of freedom aspect of the point (x, y, z, t) in its frame of reference.

This completes the derivation of the machinary of quantum mechanics on very different assumptions from the conventional interpretation. Although no new quantum mechanical results have been derived in this section a very difference physical intepretation is placed on many of the concepts of quantum mechanics.

What has not been developed is the rotational version of this approach ie where one frame rotates relative to another, this may yield some interesting results in tachyonic physics.

Acknowledgements

The author wishes to strongly acknowledge the help of Francesca Moss of the Mathematics Dept. the University of Manchester in preparing this manuscript for which the author is very thankful. Francesca Moss isn't just good she is brilliant at her job. He also acknowledges the three physics professors and Dr G Vekstein of UMIST (which has now been fully integrated into the University of Manchester who supervised and examined his work namely Prof.M.G.Rusbridge, Prof.Jan Hugill and Prof.P.K. Browning in another area of physics,Plasma Physics;they taught the importance of making sure the mathematics is right and being patient in uncovering errors. You can have an intuitive insight in physics but unless you back it with some proper mathematics and equations then it is difficult for other physicists to accept or reject your ideas. This is what Professor Browning,G.Vekstein and Jan Hugill taught the author. From Prof. Michael G Rusbridge the author learned many things about the joy of scientific discovery and the work required to confirm its accuracy. Although the key results in the Rusbridge-Sutcliffe drift wave launching theory are contained in the 1992 MSc thesis by the author which was prepared with a lot of help from both Francesca Moss and Mrs S Calland who also typed part of the PhD work it was only in published in 1997 in the journal of plasma physics and controlled fusion after detailed experimental testing and checking. It is one of only 5-10 experimentally tested or indeed testable theories concerning the launching of plasma waves in the plasma physics branch of physics that actually works. Ie it gets the predictions close, something very difficult to do in plasma physics.(Dr John Elliot and Dr John Sandeman carried out the painstaking work of doing the experiments)and in so doing were able to exceed the earlier scientific work of a former head of General Dynamics in the USA who had not

91

created a predictive experimentally testable theory on his earlier work on plasma drift waves but was keen to tell us of his work sending a letter from America for which the author is grateful ie.its nice when Americans acknowledge that Manchester is one of the centres of world science, especially since the founders of IBM obtained some of their ideas and subsequent patents from the Computer science department at Manchester University. The idea of intellectual property isn't something the founders of IBM were particularly respectful of in the late 1940s.But this is a digression.

The author also strongly feels, the importance of the Russian Academy of Sciences in the 1990s was very important to this work. The freeing up of scientific discussion was very important to this work allowing ideas that were and to some extent still are revolutionary to be openly discussed at international conferences on physics. The author acknowledges the work of Dr M.P Varin, (also Professor Dennisov whose 1998 publication entitled myths of relativity enabled different view points to be expressed freely) and other Professors such as Professor Parshin of Leningrad University and Professor S.Grigorian of Moscow University who were prepared to openly discuss new thinking in the world of physics challenging existing orthodoxies. If the author had not had that opportunity to present his ideas at these scientific conferences, and get the ideas on the scientific record at least in part, this book would not exist. The author acknowledges those who hosted him especially Dr M. Kaimakov Lina Kazakova and Professor of heart surgery German Sokorenko all of whom were very personally kind. The author acknowledges the interest of Mr A.H Bell.BDS,Prof Wendy K Olsen and J.Dahl in the completion of this work. The author acknowledges the assistance of Dan Thomas, Stephanie Frame and Logan Burton of Lulu publishing. They have given me the necessary advice and assistance in bringing these ideas to a wider readership. And finally this work would not have been possible without my wife Anne, a very kind and good person.

Index

www.ingramcontent.com/pod-product-compliance
Lightning Source LLC
Chambersburg PA
CBHW081053170526
45165CB00006B/2263